Why Systems Fail

Why Systems Fail

And How to Make Sure Yours Doesn't

David A. Turbide

Industrial Press Inc.
New York

Library of Congress Cataloging-in-Publication Data

Turbide, David A.
Why systems fail: and how to make sure yours doesn't / by David A. Turbide.—1st ed.
224 p. 15.6 × 23.5 cm.
ISBN 0-8311-3059-8
1. Manufacturing resource planning—Data processing. 2. System failures (Engineering). 3. I. Title.
TS176.T873 1996
658.4′038′011—dc20 95-44985
CIP

INDUSTRIAL PRESS INC.
200 Madison Avenue
New York, New York 10016-4078

First Printing

WHY SYSTEMS FAIL

First Edition

 Composition by Santype International Limited, Salisbury, Wilts, U.K.

10 9 8 7 6 5 4 3 2 1

To Tom Glaza

—for his encouragement and support through all these years.
A good and true friend.

Tom was a pioneer in the MRP II movement and has consistently dedicated himself to keeping the focus on what is really important—generating bottom-line business benefits through a well-planned and -managed implementation and lots of user education.

Acknowledgments

My thanks to Woody Chapman at Industrial Press for his patience and encouragement. This book took two years to write, twice as long as either of us expected, and Woody has never expressed any disappointment or concern.

Thanks also to my wife and partner Debby, and my beautiful daughters Danielle and Darcie for being there for me.

Over the last ten years, I have written more than 160 magazine articles and columns, mostly dealing with systems and implementation. Many of the ideas, experiences, and opinions expressed in these pages first saw the light of day (and printer's ink) in these publications. Foremost on the list is a magazine now called *3X/400 Systems Management* for which I wrote a monthly column titled "Midrange Manufacturing" for a number of years. I wish to acknowledge the support and encouragement of two "Systems" editors, in particular Ann Hedin, who gave me my first "break" in the magazine-writing business, and Renee Robbins, who encouraged me to stretch beyond my self-imposed limits. Both were truly a pleasure to work with.

Contents

Preface

I really don't like to talk about failure, and I apologize up front for the negative tone in the title of this book. I'm a "glass is half full" kind of guy. If there is any possibility of making something good come out of a tough situation, that's what I'm pulling for. My purpose, therefore, is not to emphasize failure, but to illustrate and discuss those factors that are typically responsible for disappointing results. By learning from the mistakes of others, we can sometimes avoid making those same mistakes ourselves. I hope there are some things in here that will help you either avoid these common mistakes or help dig your way out of the ones you are now facing.

In any case, when discussing a system implementation, it is seldom a case of success or failure in black-and-white terms: success meaning market dominance and all principals become millionaires, and failure meaning we're all looking for a job. Failure of an implementation doesn't usually result in the company going out of business. Neither does success ensure unlimited profits and market dominance. Information systems are tools—nothing more. Properly applied, information systems can certainly contribute to a company's success (however you care to define that), or, improperly implemented, they can be a burden that can contribute to a marginal company's ultimate demise.

Failure, then, is really failure to achieve the full measure of benefits that implementing the system could have provided—failure to live up to the project's full potential.

But, don't blame the system for failure or credit it for success. It is merely a tool—one of many that a company may employ in the battle against the competition, against inefficiency, against waste, against time.

Computers have become ubiquitous in today's business world. Most companies, even the "Mom-and-Pop" outfit down the street, use PCs or larger systems for basic accounting, decision support (spreadsheets, database programs), to manage mailing lists, track inventory, and on and on.

It has been my observation, however, that just about every computer system in use today is underutilized. Even systems that are overloaded, "maxed out" as they say in the vernacular, could be better applied, more efficiently operated—more useful.

It has also been my observation, based on more than fifteen years in the systems implementation business, that implementation failure is almost never the fault of the system (hardware/software) itself. We will spend some time together, in an early chapter, discussing the system selection process and it *is* important, but improper selection in and of itself is generally not the root cause of implementation failure. I have seen companies make less than optimum choices yet be quite successful (admirable bottom-line results) whereas others have the best system for the task yet fail to achieve acceptable results.

The difference? That's what this book is all about.

My experience falls mostly in the area of manufacturing companies and Manufacturing Resource Planning (MRP II or, more simply, MRP) systems.[1] What I have observed, however, applies to all kinds of businesses and all kinds of computer systems. MRP is a good example to look at because it is an integrated approach that brings virtually all areas of the business together into a single application set. MRP is one of the more integrated and complex applications, so, many of the issues to be addressed are clearly evident in the MRP example. What is true about MRP implementation is generally true about every other kind of system implementation.

---

[1] MRP is generally used to denote Material Requirements Planning, and MRP II for the more all-encompassing Manufacturing Resource Planning. In this text, I use MRP in the generic sense to mean MRP and MRP II and all variations thereof, including Enterprise Resource Planning (ERP), Customer-Oriented Manufacturing Management Systems (COMMS), and others.

For those readers who may not be familiar with MRP or MRP II, there is a brief outline in Appendix A that will give you enough background to understand better what the people in the examples in this book are dealing with. If you have no MRP experience or would like a quick review to get you started, read Appendix A first.

I expect that many readers will say, "that's us!" at one or more places in this book. This happens to me all of the time. Whenever I make a presentation that touches on problem areas, and especially when I use exaggeration to make a point, people come up to me at a break or at the end of the session and say, "When were you at our plant?" or "You were talking about us, weren't you?" Rest assured that the people and the situations you read about here are disguised, and in nearly every case, represent more than one company's experiences. You can also be quite confident that if you see yourself here, you are definitely not alone.

We, as imperfect human beings, tend to make the same mistakes over and over. It's sad, but we also tend not to learn from the mistakes of others. There is no rocket science here. Implementing an information system does not require the future users of the system to become programmers, computer scientists, or systems analysts. If the system does require this kind of expertise, there is something radically wrong with the system itself. You may choose to become more knowledgeable about what's going on "under the hood," but you shouldn't have to know the technical details any more than you would have to know how an internal combustion engine works to be able to drive a car.

Consequently, you will not find programming tips, technical system details, or computer jargon in this text. There is one chapter that deals with the system selection process, but that is as technical as it's gong to get. It is not necessary to know anything at all about computers to understand the kinds of problems that cause systems to fail—they are all people-related. But, I would bet that you already suspected that, didn't you?

Understanding the function of the system, and I use the word system in a very broad sense, is certainly necessary to get the most out of it. Whether the system is computer hardware and software, or manual procedures, or (more likely) a combination of the two, the user should know what he or she is doing and why it is to be done in that way. I firmly believe that the native intelligence of employees is most often severely underrated and underappreciated. Managers must learn to trust and respect that intelligence and be willing to invest in *education* for employees as well as *training*. More about this later.

Why Systems Fail

1. Introduction

A company has a need. Inventory is out of control. Production schedules are in disarray. Shortages abound. Customers are calling to find out where their (past-due) orders are. Everyone agrees that the solution is to install a new information management system.

The company hires a consulting firm to prepare a request for proposals (RFP) that it duly distributes to a dozen or so vendors. Several weeks later, the vendor responses roll in, the consultant applies a complicated formula to rank the responses and the two or three companies at the top of the list are invited to demonstrate their solutions.

A selection is made, the vendor moves in, and implementation begins. Six months later, the CEO calls for a meeting of the company's top management team with the system vendor.

The vendor knows there will be fireworks. The project justification called for substantial benefits by this time and they are nowhere to be found. The meeting goes about as expected. There is a great deal of shouting. Accusations, and excuses fly in all directions, threats are made, and tempers boil.

The end result of the meeting: The vendor makes more promises that cannot be kept, company management has even less confidence in the vendor and the system than before, and everyone starts building up a protective wall in hopes of avoiding the flying shrapnel when the whole project eventually blows up, as everyone is sure that it will.

At the end of a year, or maybe eighteen months or so, the vendor is gone, a few features of the system are in use, maybe, and the business is still out of control. Some of the best and brightest employees have left in frustration or found more promising venues. Those responsible for the project have been fired or demoted. Those who did not bail out at the first signs of trouble, that is. The naysayers who "knew all along" that it wouldn't work are praised and promoted. And the "system" takes a good share of the blame.

The new management team goes through the process again the following year. A new system and new consultants are selected, the project is justified, planned, and undertaken. This time the goals are more modest and measurable benefits result. It is generally agreed that the second implementation was a success and the first one failed because the wrong system was installed using the wrong vendor and consultants.

Was it poor system selection that prevented success the first time around? Were the vendor and consultants really to blame? Perhaps, but I'd be willing to bet that the results could have been significantly better, with the same system, if a number of things had been done differently.

Among that list are many that involve people—the employees of the company, management, the project team that manages the implementation, and the outside support resources. There are a few that involve computer hardware and software, but those items are at the bottom of the priority list.

I have been a manufacturing systems consultant since 1981 and I have seen this process, system selection and implementation, hundreds of times. I have been involved as a consultant writing the specifications and preparing the RFP, as a vendor technical support person responding to these requests, and more often as a postsale implementation resource. Many times I have been called in months or years after the initial implementation to help a management team to reimplement or extend the use of the installed system. In these roles, I am often asked what are the most important "keys" or "secrets" to successful implementation. There are only three.

The first is full and active commitment from senior management. Without this, the project may get started but will surely bog down at some later time and fail to achieve its objectives. The second key is organizational. The project must be planned and organized by a cross-functional team, management must be prepared to deal with the changes that will undoubtedly take place within the organization to take advantage of the new system, and the organization itself will have to adapt to a new dynamic as the system's impact starts to take hold. The third key is educational. Education and training are the weapons that can help overcome fear of technology, fear of the unknown, and the natural human reluctance to change. This will be the best investment that you will make in your implementation budget.

The stories and advice you will find in the remaining chapters of this book expand on these few simple truths.

The real tragedy of system implementations is the unrealized potential and the fact that, in most cases, the difference between the available benefits and the realized benefits is often a few simple errors of omissions in the implementation process. And worse, we repeat the same mistakes time and again. When I was growing up, my dad used to tell me that I should learn from other people's mistakes because I would never live long enough to make them all myself. I believe that this is good advice and I have made a career of learning from my own mistakes as well as those of others and passing on this wisdom.

This, in fact, is probably the biggest benefit available from outside consultants. Consultants who have "been there" and observed the good and the bad examples of systems implementations should be able to steer a company in the direction of the former while avoiding some of the mistakes of the latter. Most companies will only implement a new information system once every five to ten years, whereas a consultant may experience a dozen or more different implementations in a year.

I believe it was Mark Twain who said, "Good decisions come from experience and experience comes from making bad decisions." Human beings make mistakes. It's only a tragedy when we insist on repeating them.

What Is Success?

Since the overall theme of this book is success and failure, it might be valuable to spend a little time looking at these classifications and coming to an understanding of what they might mean.

Success is not an absolute or easily identifiable state. It is both subjective and relative and depends to a great extent on expectations. Perhaps surprisingly, it is often difficult to recognize success simply because it isn't defined, and each observer is allowed to apply his or her own criteria.

You will find numerous references in later chapters to goals and measurements precisely because they are the foundation of the definition of success and the tools for an objective view of it. It is very important to set these (measurement) tools in place before the implementation is started.

Often, a financial justification is required to get authorization to buy the system hardware and software. This usually takes the form of a cost/benefit analysis using traditional discounted cash flow techniques. The project champion or initial project team will gather cost information, estimate the value of direct benefits, and compare the two to determine the return on investment (ROI). The comparison considers the time value of money that factors in the effects of inflation and cost of borrowing or not investing the cash required to implement the system. Typically, there is a company policy (perhaps unwritten but almost always in effect) that determines the required return on the investment necessary to justify the project.

Let's look at a simplified example. Let's say that a system will cost $1 million, which is to be paid up front. (I said this is a simplified example. The total cost is never paid all at once and will seldom be a nice round number such as I'm using here.) The projected benefits are as follows:

Year 1	$100,000
Year 2	$300,000
Year 3	$400,000
Year 4	$400,000
Year 5	$400,000
Total:	$1,600,000

Looks good, doesn't it? Invest a million, get 1.6 million back. But let's assume that inflation is projected to be a constant 4% per year through this time frame. Discounting the value of the benefits by the inflation rate, the revised chart looks like this:

Year 1	$100,000
Year 2	$288,000
Year 3	$368,640
Year 4	$353,894
Year 5	$339,739
Total:	$1,450,273

The total return, then, over a five-year period, is the amount shown. Converting this to a percentage of the initial investment and dividing by five (years) give us a (simple interest) annual return of 9%. On a compounded basis, the return is only 7.7%. If the company has the cash, it might be already invested in some kind of securities that return more than this. If the money must be borrowed, the percentage of interest to be paid might be higher than 7.7%. In many companies, there is an informal expectation that any capital investment must return at least a certain minimum percentage (often 15% or more). This project might also be competing for funds with other projects that have higher returns.

So, initial justification usually includes a cost/benefit financial analysis that can be a crude criterion for success. This analysis is frequently based on direct return—things like reduced payroll cost (labor savings), higher output per dollar of labor (efficiency or productivity), or reduced inventory investment. Often, however, the biggest benefits of a system implementation are the indirect or intangible benefits including such things as better communications and coordination between departments, more timely information on which to base decisions, and customer satisfaction (which can lead to more business) due to higher quality, better responsiveness, or shorter lead times.

Just looking at the cost/benefit analysis, though, can you make a success/failure determination based simply on whether the specified return was actually achieved? If the justification projected a 15% ROI and the actual return was only 13% was the project a failure? What if the return was 9%? Can it be considered an unqualified success if the return was 15.00%? What if the real potential benefit was a 50% ROI, but the justification only specified 15%, and the project actually achieved 18%? This project beat the justification but left a lot of potential benefits on the table. Even using a simple cost justification as a measure of success leaves a lot of room for interpretation.

Success will have a different meaning if the new system is a replacement for an existing one, as opposed to a new facility altogether. A company may have a "good" MRP installation yet decide that it should upgrade to a newer or bigger or better system, or simply a later version of the existing software. In this case, most if not all of the direct benefits to be derived from MRP II are probably already in hand. The justification, therefore, may be based on high support costs for the old system or limitations of the incumbent system that would not support the company's growth. Success in this situation might be simply having the replacement system working to the same level as the old one—a direct replacement. This is the only case where I can envision the measure of success as simply having completed installation. I would hope, though, that the company would also be interested in something more: some new capabilities or improved management that would result from the newer, better faster system.

Systems, whether new or replacements, must be able to pay for themselves by providing worthwhile benefits. If the "something more" is not clearly defined before the implementation begins, how will you know if you achieved the desired result? It is far too easy to make the target match the results (shoot the arrow first then draw the bull's-eye around where it lands) if not set before implementation.

Targets (goal) can be either absolute or relative. Relative targets are much better. Let's say a company has $10 million of inventory. One goal of the implementation project might be to reduce inventory to $5 million in two years. Ignoring the fact that this is not specific enough (what kind(s) of inventory) or appropriately stated (more about setting goals and measurements later), the company could dig in and achieve this goal. Obviously, this would be a benefit, but is it good? Is it the best that could be done? Does it make this company competitive?

If all of the other companies in the same line of business have inventory (relative to equivalent sales volume) higher than the $5 million level, then there wouldn't be any question that the project was a success. Even if the competition is still ahead, though, achieving a 50% reduction in inventory can be hardly considered a failure, can it? What happens next, though? Should they sit on their laurels, content to maintain the $5 million inventory level? One problem with absolute goals like this is that they are static. You achieve the objective or you don't. End of story.

A relative goal for this same situation can say the same thing and more. Let's state the same goal as a 50% reduction in inventory, but time phase it and make it continuous. For example, the target could be a 30% reduction in twelve months and a further 20% reduction the next year, then an additional 10% per year beyond that. It will take a few years to achieve the 50% just as in the other example, but the story doesn't end there. Each year, this company strives to be 10% better, making the results not only higher than the simple 50%, but continuously improving beyond that.

Realistically, this target is not adequate either, unless the company's sales volume and market situation are not changing (a very unlikely situation). An inventory reduction goal should be stated as a ratio to business volume. A company may start out with $10 million in inventory to support $20 million in sales. If sales volume is increasing by 15% per year, it may be an aggressive goal to simply maintain the inventory at $10 million because the increasing volume makes the $10 million inventory a smaller and smaller fraction each year:

	Sales	Inventory	Inventory/ Sales	Improvement
Year 1	$20M	$10M	50%	—
Year 2	$23M	$10M	43%	14%
Year 3	$26.5M	$10M	38%	12%
Year 4	$30.4M	$10M	33%	13%

So, measurements are not as simple as they might appear, and they can be misleading. The problem with inappropriate or misleading measurements is that they may drive the wrong behavior or the wrong results, or they may hide benefits, ignore problems, or simply not measure the impact of what you are changing.

Benchmarking

One of the more popular buzzwords circulating around in the last few years is benchmarking—often associated with reengineering and/or quality improvement programs. The idea behind benchmarking is that a company should identify the "best" in their business or market and measure what it is they do well and how they do it. Using this competitive standard or benchmark as a guide, the company should then emulate the best practices of these industry leaders. Benchmarking can provide perspective for setting goals and measurements.

One problem with benchmarking is that it can become a project in itself, and an expensive and time-consuming one as well. In some cases, the benchmarking process turns into a series of field trips that everyone enjoys but from which few real benefits result.

Nonsystem Benefits

In addition to the improvements that can be directly attributed to the services provided by the new system, there are improvements to be gained from process changes that occur during the project. In truth, it is usually difficult to separate one from the other and there is little or no reason to do so. The point is, however, that the majority of the benefits result from improved procedures, which are supported by better availability of information. Success (benefits) relies more on how well the people and the organization manage the business using the information, not simply having the information. Stated like this, it seems all too obvious. Keeping this all in perspective, however, is difficult when the majority of the visible project costs are computer hardware and software. We expect a return on that investment and when we don't get it we blame the system.

Why Systems?

Setting the direct and objective measurements like inventory reduction aside for the moment, there is something to be said for better management, better control, and better communications. Even though these may not be quantifiable results of a system implementation, they can be among the more important outcomes.

What does an information system do? Simply, it manages information. The input to the process is data—facts that, in themselves, have little meaning. Data include definitions, conditions, or events (facts, figures, measurements). When combined and presented to the user in useful formats, data become information. The word information is based on the word "inform," which, according to Webster's, means "to give form to; to give life to." So, changing data into information gives them life and, I dare say, value.

Continuing on this same scale, information can be the basis of knowledge (clear perception, learning) that leads to wisdom (sound judgment, sagacity).

Data → Information → Knowledge → Wisdom

A computer system participates only on the lower end of this continuum, receiving data from sensors or, more often, human input, and through its ability to store, sort, and combine the data produces information. It is up to human users to advance the information to the next rung of the ladder by interpreting the information and putting it into context. We learn from the information and use it to make decisions. Whether the information becomes knowledge or leads to wisdom is totally out of the control of the computer.

Computers do not make decisions nor do they manage anything but information. Much has been written about artificial intelligence, expert systems, neural networks, and the like, but none of these is anything more than the application of human-defined rules and processes and an extension of information management mechanisms. Computers do not exhibit judgment and computers cannot be wise.

Assessing system implementation success, then, really has little to do with information system function. What we should be measuring is how well the people in the organization have been able to apply their judgment and wisdom in managing the company to achieve the desired results. The system is merely a tool to make information available in a form and a manner that supports the needs of the users. Implementation of the information system is both the excuse for the changes in the way business is conducted and the underlying facility to be used to manage the information we will use for management improvement.

Success

We will be looking at many different varieties of success and failure in the following chapters. Keep in mind, throughout this journey, that success is relative and quite subjective. Any improvement in the management of a company, whether direct and measurable or entirely intangible, can be considered a kind of success. When companies spend money, time, and effort implementing a system, there is a certain expectation of return on that investment. If that expectation is not met, the project could be considered a failure even if substantial benefits are achieved. In all cases, however, there is always more that could have been achieved. No matter how successful the project, therefore, some benefits will be left "on the table" and that might be someone's idea of failure.

Without being overly judgmental, then, let's look at some fictitious situations and discuss what went wrong and how these errors could have been avoided.

2. Packaged Software

Selection, Modification, and Use

Two companies, Acme and Zebra, installed similar MRP II systems three years ago. Both failed to achieve the benefits that they had anticipated when the implementations began. It happened that both companies decided to try again at about the same time.

Acme, understandably perhaps, was disappointed in the chosen software package, so it prepared an RFP and went shopping for a replacement. A suitable candidate was chosen and eventually installed in place of the old, failed system. Three years later, it was ready to throw out the second package and try yet another.

Zebra, on the other hand, decided to retry with the original package. There was considerable resistance to this idea because the system was, of course, closely associated with the failure. Management prevailed, however, and the project was restarted. Three years later, Zebra was a "Class A"[1] MRP II installation and the users were completely satisfied with the system and the software package.

Acme blamed the initial failure on the system, ignoring the real causes of the failure, which were all people issues: lack of senior management involvement, no strong project team, and insufficient user edu-

---

[1] The Oliver Wight organization promotes a rating system for the effectiveness of an MRP II implementation based on the response to a series of questions that rate how well and how completely the system is used. Class A is the highest grade. *ABC Checklist.*

cation and training. When it came time to try again, it switched systems but, having failed to recognize the real problems, it was destined to repeat them—with the same results.

Zebra, on the other hand, realized that the system was not at fault. Armed with this nugget of knowledge, it uncovered the real problems and addressed them in the reimplementation effort. Not only was this second effort successful, yielding the benefits that were not gained the first time around, but it did so with a relatively small additional investment.

It's easy to blame the system. Far too easy, in fact. The system can't defend itself. It is just sitting there in all its high-tech glory with its monthly bills to remind us of its presence. It is the most obvious physical evidence of the implementation project.

Reimplementation, however, is not easy. Whatever the real reason for failure the first time around, the system will be seen as "no good." Unless you can get past that perception, any reimplementation will fail because there will be no confidence in the system, which will lead to an unwillingness to fully participate in another try.

As a result, reimplementation usually means restarting with a new software package. If the proper lessons were learned the first time, the second effort may well be successful and the impression that the first software package was at fault will be reinforced. Oh, well, I guess what really matters is that the implementation is finally successful, not who or what is blamed for the first failure.

With any software package, it is possible to find good references and bad ones—successes and failures. When shopping for a system, if you ask the vendor for references, of course it will provide successes. You may also become aware of failures through your own contacts or other means (competitive vendors might steer you toward failures, although I'm not sure that's very ethical). Be careful about drawing conclusions from either successful examples or failures. There is much more to the implementation than software fit and functionality.

Application software is a very competitive business. There are many vendors in every segment of the market. Software is a labor-intensive business with little capitol equipment required, therefore, it is an easy business to enter. I could spend a few weeks of my spare time writing some programs and instantly become a packaged software vendor. You will want to be sure that you are dealing with established, reliable vendors, but that is a topic for another chapter.

Among the established vendors in any market segment, competitive advantages are extremely fleeting. One vendor might release an update to its package that, for example, includes an Executive Information System (EIS),[2] the first such facility available in their market. Before long, however, all the significant competitors will add an EIS to their systems. Thus, packaged offerings grow, and continue to be alike no matter how much they change. The exception to this rule is when a vendor focuses on a market niche rather than adding features designed to reach a broader market. Here, too, there are likely to be competitors within the niche who will copy significant features of the leading software packages in that niche.

Isn't this copying illegal? Not necessarily. Although there are exceptions, added features usually cannot be protected by patent or copyright. When there are infringement suits, they take years to resolve and it's hard to predict which way the courts will go.

In any case, packaged software products in any defined market tend to be far more alike than different. Finding the most suitable product for your company is often a process of splitting hairs. To be honest, software (system) selection is hardly ever based on functionality, although that is usually the announced reason. The real selection is most often based primarily on the financial strength of the vendor, personalities of the sales representatives, promises of support, price, and similar reasons. And that is as it should be.

No packaged software product will be a 100% perfect fit in any particular situation. It is likely, though, that there will be several (at least) packages that will fall within a narrow range of each other. Maybe they'll range between 75% and 85% fit (however you choose to measure fit). There will be minor differences between them, of course, but none so important that it would swing the decision one way or another. Selecting from among the finalists cannot be based on functional differences.

The degree of success or failure of the implementation cannot be attributed to such minor differences in function. Choosing the 40% package rather than the 80% package could be disastrous. This,

---

[2] An EIS is an advanced data retrieval tool that presents summarized, high-level information in a graphical format. The type of information is predefined to include the kinds of things the executive likes to see on a regular basis.

however, is a highly unlikely event if any attention whatsoever is paid to the selection process.

Packaged Application Software

I'm working from the assumption that the system being implemented includes a packaged software product. Today, this is the case more often than not, but just about everything that follows in this book applies to custom application software as well as packages. It is appropriate, however, to deal with the software design or selection process at this early stage so that these concerns won't interfere with subsequent topics, which are all people, project, and management concerns.

I am an unabashed supporter of packaged application software. I also believe that the industry is changing and "packaged" software may be quite different in a few years. Appendix B includes a reprint of a magazine article I wrote in late 1993 that hints at some of these potential upcoming changes. In any case, the economic argument for packaged software is compelling: A package can provide more function for far less cost than customdesigned and written programs.

Because there is no such thing as a free lunch, of course there is a downside to packaged software. As with any other off-the-shelf product, you don't get to design it. Thus, packaged applications seldom, if ever, provide exactly the function you want—there will be more function than you want in some areas and less than you need in others.

It's not actually as bad as that may sound. A well-designed package will include all of the major functions required to address a particular business or business area and probably more than any particular company can use effectively, at least at first. The buyer usually ends up with more function than they need at far less cost than would have been required to build a (more limited) custom solution. A bigger problem is the perception of uniqueness.

I remember walking into a company in New Hampshire a few years ago and having an initial meeting with Jim, the materials manager, in a quiet corner of the plant's lunchroom. Jim introduced me to the company organization, its products, and customers. He then outlined the problems that I was there to help the company address. At one point during this conversation, Jim looked me in the eye, smiled, and said, "You know, we really aren't any different than anybody else." That was the only time in the last fifteen years, in my dealings with literally hundreds of companies, that anyone has ever told me that.

At least in the United States, and in much of the rest of the world as well, everyone is an individual and proud of it. This attitude carries over to the organizations that we work for. We are forever trying to distinguish our products, our processes and quality, our customer service, or whatever it takes to beat our competitors in what have become very crowded markets. My observation, however, is that procedures and information systems needs aren't really that much different from one company to another.

Even in manufacturing companies, where products and processes span a staggering range, there is a lot of commonality. Whether a company makes cars or can openers, soup or super computers, pencils or precision machinery, it buys materials (or parts), brings those parts into a process that changes them into products, and it distributes the products in some fashion. These sound like grossly general descriptions, and they are, but the information management needs of all these environments includes purchasing the parts or materials, incoming inventory handling, issuing materials to the process, managing the process itself, moving the completed products to finished goods inventory, then through customer orders and shipping. The process can be described, scheduled, tracked, and costed using definitions of the facilities and their capacities. Planning applications can simulate all of this activity based on definitions of the products and processes, and demand or forecast information.

The terminology is different in different industries: What hard-goods manufacturers call a bill-of-material is usually called a formula in a chemical plant or a recipe in food processing. And the emphasis is different. Some companies have complex processes with few materials; some processes are simple, but take a lot of time; other companies face complex material planning and management needs. In any case, the differences are more a matter of emphasis than of fundamental, structural uniqueness.[3]

A general-purpose package, therefore, should address the basic needs in as flexible a way as practical. MRP II packages started out

---

[3] See *MRP+* (David A. Turbide, Industrial Press, New York, 1993) for more discussion of differences in the use of MRP in various industrial environments.

with a "typical" manufacturer in mind. Typical could be defined as a company that fabricates and/or assembles products from purchased components and materials in convenient (economic) lot quantities. The end products are stored in inventory and sold from this stock. The manufacturing process includes a number of process steps that vary by product and can be defined ahead of time.

After initial market acceptance and the emergence of competition among MRP software package vendors, however, a broadening of applicability has become the hallmark of the industry. Features were added to address the needs of custom manufacturing (make to order rather than sell from stock), process industries such as chemicals and pharmaceuticals, and high-volume repetitive manufacturing. New functions support bar-code data collection, electronic data interchange, and maintenance management. The trend continues. Newly announced enhancements include configurators (for engineer-to-order products), field service management facilities, quality tracking, marketing support, enhanced distribution management, and others.

The good news is that most packaged MRP II systems are modular, that is, you can buy and install individual application modules and not buy applications that you don't need. This is a bonus for future flexibility because you can add modules later if your business needs change. Most packaged systems also offer some flexibility in how they are installed and used, including the ability to "tailor" the function and the appearance of screens and reports.

Nevertheless, buying and installing a package are quite different from custom designing and building your own software. A custom package will (conceptually) do exactly what you want it to do in exactly the way you want it done. With a package, you get what the vendor provides. Of course, you can always modify a package (more about modifying a package later in this chapter).

With both custom and packaged code,[4] the glass can be either half empty or half full. The positive side of a custom code is that it will be uniquely suited to your needs. On the other hand, it will also institutionalize current procedures, whether they are the best way to do things or not. Custom code will also reflect your needs and the

[4] Please forgive my tendency to refer to computer programs as "code." To me, "program" as a noun and "code" are synonyms. Code can also be used as a verb, meaning the act of programming. I promise not to use "code" that way.

way you do business today. By the time the programming task is complete, those needs may have changed and the software function will lag behind changes in your business. Changing specifications (during development) are the bane of the programmer's existence and probably the largest cause of cost overruns, schedule slips, and dysfunctional code. Finally, writing programs does take time and any benefits to be derived from use of the system will be delayed during the development cycle.

Packages are inexpensive (relative to custom code), and are available immediately. The problem is the imprecise "fit" between package function and perceived need. This can be an advantage, however, in that the package is probably designed according to either "industry practice" or "standard" ways of doing things and probably contains more features and functions than you are currently using. By adopting the package's procedures, you may be able to move up to better management practices. A counterargument could be that your company is successful *because* you do things differently.

Now, after trying very hard to be fair to both approaches, I will let my prejudice surface and discuss both the argument for packaged software and the process of selecting and installing same. Much of the remainder of this chapter is adapted from a similar discussion that appeared in my book *MRP+* (Industrial Press, 1993) and covers selection of a package, living with differences, and adapting or modifying the package.

After this chapter, I will not discuss software selection or system functionality. I will assume that the software does what it does, that its function is appropriate for the situation, and that any software limitations are addressed as discussed in this chapter.

When a new system is installed and the result is less than expected, the easiest target for blame is the system itself. That way, nobody has to take responsibility except perhaps those who chose the system in the first place. As you will see in later chapters, there are many other more important causes for failure (lack of success).

Selecting a Packaged Solution

It is generally recognized that there is no such thing as a perfect fit of packaged software to a company's specific needs. Most people would find any package that satisfies more than 80% of the requirements is a good solution, although I'm not sure how you would measure this percentage of fit. Does packaged software make sense, then? Absolutely.

Packaged software offers many advantages over custom-written software, not the least of which are low cost, proven function, availability of education and support services, and ready availability. To develop your own custom applications today is almost always too expensive, takes too long, and offers significantly less capability than a packaged solution, despite the less-than-perfect fit.

Ironically, custom software is hardly ever a perfect fit either. Because of the practical limitations of time, budget, and capability, custom applications cannot be as complete or comprehensive as a packaged product. During the extended development cycle, the company's needs will change and the design of the application will either evolve with the needs or be outpaced by them. If the software specifications evolve during development, the initial design will almost certainly prove inadequate and either the function will be compromised because of these limitations or costly redesign will delay the project further, making the suitability of the design subject to further erosion.

Changing specifications are a software developer's nightmare. This is the most frequent cause of dissatisfaction and rancor between the developer and the customer (user), whether the development is being done by an outside contractor or your own data processing staff.

What are the options, then, when faced with the prospect of selecting and implementing a less-than-perfect package? The first challenge is to find the package with the best fit. Then, learn to live within the limitations of the package, adapt the standard offering to better serve your needs, or you may choose to modify it.

There are literally thousands of packaged application software solutions available today. They range in price from a few dollars to more than a million and are marketed by computer hardware companies, general software companies that have packages for many applications, manufacturing (or other) software specialists, systems integration companies, accounting firms, consulting firms, the computer store at the mall, and the whiz kid down the block. Choosing the right package can be very challenging.

The Request for Proposals

To find the best fit, you must first understand your needs. Some companies, especially larger ones with deep pockets, will commission an outside consulting firm or accounting firm to prepare a formal request for proposal (RFP) that can be sent to a number of vendors, soliciting formal presentations of their solutions. Having been on both sides of this approach, as a consultant preparing RFPs and as a vendor responding to them, I have mixed feelings about the validity of the RFP approach. On the one hand, it is essential to analyze and document your needs, and communicating these needs effectively to the vendors gives them a good opportunity to respond with a valid proposal. On the other hand, however, this process is often overdone. I've seen companies spend more on the analysis and RFP process than it would have cost to buy and install a very good mid-range solution.

Some of the larger accounting firms will charge many tens of thousands of dollars for this service, and will prepare a document that is made up mostly of "canned" questions that may or may not focus on the critical issues. Often these "professional" RFPs will contain several thousand questions, most of which describe in minute details the features and functions that virtually every MRP II system worthy of the name will include as a matter of course. Worse yet, these questions are often what the vendors call "wired," that is, they describe the details of one particular package, making it difficult for any other package to look good by virtue of having too many "no" responses. This wired nature is sometimes inadvertent, as a consultant who is most familiar with one particular package will include its features as minimum requirements and often will not know details about features of other packages that may be equally or more important for this company. The problem is magnified when the preparer firm is associated with the sale and/or installation of a package of its own.

The problem on the respondent's end is that these extremely detailed RFPs are often received with very short deadlines for response. I have on a number of occasions been asked to respond to a questionnaire of more than one hundred pages, with over a thousand detailed questions, in just one or two days. The RFP may well have been sent out with a month to respond, but the delay in delivery, routing to the proper sales rep, initial assessment (do we want to respond?), assembling the right technical team, etc., can easily use up most of the time.

When there are a thousand or more questions, the bidder cannot afford to do more than cursory analysis and the responses are often limited to the following: (1) we comply with this requirement; (2) we don't do it quite this way, but we do something similar; (3) it's not a standard feature, but we can adapt (modify); or (4) no, we don't have this. Writing the detailed explanations for answers (2) and (3) is a challenge, especially when done on a very short deadline. In addition, there is limited opportunity for the vendor to promote the unique advantages of her product within the relatively rigid framework of the RFP questions.

The evaluation of these proposals is often done with a point system for scoring. The number of positive responses, or qualified yes's, and no's, are tallied, and usually weighted to reflect the relative importance of each feature. There is little room for judgment within this format—the item is usually either satisfied or not. Total weighted scores are used to reduce the number of contenders to a reasonable number, usually two to five. The finalists are then scheduled for presentations and demonstrations before a final selection is made. The entire process takes a minimum of three to four months, often longer.

The RFP, however, can be made to be a useful approach to system selection without being overly complex or expensive. Rather than hundreds of pages of marginally useful, detailed "boiler plate" questions, the RFP format could be used to describe the business needs of the company and focus on the primary areas of concern. It should describe the current situation, primary concerns (problems), and the goals and direction of the company for the next three to five years. It should invite the bidder to visit the plant and talk to the prospective users, all the better to understand the company and its situation.[5]

In addition to functions and features of the software, the RFP should seek information about support services and implementation assistance. What do the bidders have to offer, and what other sources are there for support?

You will also want to know about the vendor(s) you will be dealing with. How big are they? How long have they been in business and how stable are they? Insist on references and be sure to

[5] The author has developed a streamlined system selection process called "The Lean RFP," which is documented in a white paper report available from Production Solutions, Inc., Beverly, MA.

talk to some of them. Try to find satisfied users of similar size with similar needs.

Just because the vendor company is small should not automatically eliminate it from the competition. You must be careful, however, to protect yourself against the possible demise of the vendor. With smaller companies, check references very carefully, and insist on the use of widely supported hardware, standard computer languages, and data management utilities. Also ensure that you have access to source code so that you can find other support for your system if the company disappears.

The Invitation to Bid

A less formal approach involves the identification of a select number of vendors who are invited to visit your plant and gather their own requirements inventory. The initial selection might include companies with a strong local presence, those whose software runs on the equipment you own or are familiar with, or the largest vendors in your industry segment or hardware size category (mainframe, midrange, or micro). You will spend more time ushering the potential bidders around the shop, but you will save the up-front effort of doing the analysis yourself and writing the RFP. Sometimes several bidders can be accommodated with a group presentation and tour followed by private question-and-answer sessions.

The responses (proposals) thus generated will vary more than the answers to RFP questions and the comparison and selection will probably be more subjective, but the proposals might also make you aware of problems or opportunities that you might not have discovered on your own.

The Sales Call

Or the solution may come, unsolicited, to your door. Sales reps still make "cold calls"—through the mail, on the phone, and sometimes door-to-door. Marketing "seminars," demonstrations, and other events are held regularly to interest prospects in the products. These events are often very educational and don't usually involve high-pressure sales techniques. If you see something you like, invite the vendor to tour your facility and propose a solution.

I don't necessarily advocate that you buy a system based on one proposal from one vendor, but it's not necessarily a bad way to do business. If you have a vendor now that you are satisfied with, why not let it propose the next-generation system for you? If the vendor that sought you out has a good reputation, an appropriate solution, and you feel comfortable with the vendor, why not? As you have seen, there are accepted approaches to the big issues incorporated into many off-the-shelf packaged solutions. If the bidder's package has the right features and meets all of the other criteria, there might not be any need to look further.

So, You've Decided

Before moving on to the discussion of how to handle the unique requirements—that last 10% or 20% of the "fit"—I'd like to add a few more words of caution.

There is a lot of turnover in the software business. If you know how to program, or think you do, the only thing you need to go into business is a phone and some business cards. Every day, new software companies start in business and others disappear. Most software companies don't last. Even the big ones go broke, merge out of existence, split up, file for bankruptcy, lose their best people, discontinue products, or otherwise become less capable of supporting their customers. If your business depends on the continued operation of their software, you are vulnerable.

All software has bugs. Some of them don't show up for years. As your business grows and changes, you may begin to use parts of the package that you never used before. The error was there all along; you just didn't know it. If the vendor is gone, how will you solve the problem?

There are several ways to protect yourself. First, be sure that you have access to the source code for the programs in your package. Most vendors will sell (a license for) the source code at a modest additional charge or sometimes will include source code at no extra charge. If your selected vendor refuses, seriously consider another vendor. An alternative is to have the source code held by a third party with a contract that gives you access to the source if the vendor is unable or unwilling to provide support. This is a much less desirable solution, but is one way to protect yourself if you are not allowed free access to the source code.

An additional safety valve is to buy a package that is supported by third-party resources, that is, someone in addition to the software vendor. It is good to have a choice of several programming companies, authorized representatives, systems integrators, or other firms, independently owned and managed, that know the package and can help you support it in the event of the vendor's demise or if you decide that the vendor is not supporting you to your satisfaction. You hope it will never happen; you select your vendor carefully to help ensure that, but you can never know for sure.

Learning to Live with Limitations

It is a given that no packaged product is a perfect fit. It is hoped that the differences between your needs and the package's capabilities are minor, and you can use the package, despite its flaws, pretty much "as is." In the popular jargon, this is known as using the "vanilla." package.

If you have made an appropriate selection, the package should provide for the majority of your company's needs. There sometimes can be some differences in terminology or format between the new package and the old way of doing things, but it is highly recommended that this kind of difference be accepted and you should adapt your procedures to the package's way of doing things rather than modify the package to look like the old way (more about modifications later).

In addition to the considerations outlined in what follows, a danger in changing a package is similar to a problem with custom development: You will design for today's needs and sometimes preclude enhanced functions that you don't need today but may be able to exploit in the future. The old way of doing things obviously had some disadvantages—otherwise, there was no reason to buy the new system. If you replicate the old way with the new system, what have you gained?

Because a packaged product must be designed for a wide range of situations, the standard reports and inquiry screens will tend to be "busy," that is, will contain more information than any one user will need, in order to accommodate as many situations as possible. New users can find this annoying or confusing. Resist the temptation to remove unneeded data, if you can. It is often the case that some of those "extra" fields really are beneficial, but if they are removed, you might never uncover their usefulness.

I have seen many cases where a company has modified a package to include a "new" function when the desired capability was always there but was hidden by early modifications that removed information and capabilities that the company either did not need at the time or did not understand enough to appreciate them. In the next section, I will discuss additional reports and other add-on functions that might have the same effect as changing reports or inquiry screens. Be careful. The short-term convenience might have long-term detrimental consequences.

"Standard" Functions

Virtually every company that I have worked with over the last fifteen years has thought itself unique. The truth is, however, that most manufacturing companies do similar things in similar ways and the uniqueness tends to be more one of terminology, technique, or details. Because packaged products are designed to meet the needs of a large industry segment (all manufacturing, all process manufacturing), the functions and procedures included are pretty much the "industry standard" way of doing business. Before modifying a package, you should ask yourself why you do whatever it is you do in a way that is significantly different from the way other companies do it, as incorporated in the "vanilla" package.

I understand that companies like to feel that they have a competitive advantage because they *are* different, but often the differences we are discussing here are not significant in terms of a competitive edge. Usually, they are the result of long-standing tradition, inertia, or ego (the "not invented here" syndrome). Once you have identified the "why" behind your way of doing things, decide if your uniqueness is really a significant advantage, significant enough to justify the trouble and expense of modifying the packaged software.

Implementation experts will talk about modifying your procedures to adapt to the package rather than modifying the package to adapt to your procedures. This is by far the preferred approach for all the reasons to be discussed (modifying a package) and is usually the fastest, most rewarding approach you can take to implementation. Users will resist changing their ways. There is a natural human resistance to change. This resistance can be overcome only through adequate education in the functions and capabilities of the new package and through an understanding by the users of the advantages of the new approach.

The latter point is achieved through education, and via firm direction and commitment from above. Above all, remember that each user will approach the proposed change in his process or procedures with an attitude of "what's in it for me?" although few will actually voice this concern so directly. In order to wholeheartedly accept and support a change, a user must feel comfortable that there is adequate justification (reward) to compensate for the inconvenience, conversion, retraining, or whatever adjustments are needed to put the new function into use.

Adapting a Package

If the package is deemed to be inadequate in some significant area to the extent that a change is required, there are several degrees of change, with varying impact on the maintenance and supportability of the package.

I like to distinguish between "enhancements" and "modifications." An enhancement can be custom-developed or sometimes can be purchased from the original vendor or a third-party software supplier. An enhancement is either totally or mostly "outside" of the package. By way of contrast, a modification is a change that directly impacts the purchased program(s). This section describes enhancements. The following section deals with modifications.

Typical enhancements provide additional reports, add auxiliary functions (enhanced sales analysis, reformatted control documents, regulatory compliance reports, enhanced inquiry functions, industry-specific analyses), or extend the function of the system into niche industry segments. Modifications often change the logic of a process such as new or enhanced cost accounting functions, unique industry-specific calculations, or changes to input formats or data requirements.

If the package you chose is widely installed, chances are good that companies other than the vendor have developed additions to the package and market them as add-on enhancements.

The existence of a rich supply of enhancements is not necessarily an indication of inadequate function of the base package, but more likely reflects the diversity of the marketplace and sufficient installed base to justify development and marketing of add-ons.

Packaged enhancements offer many of the advantages of packaged software as discussed earlier: Packages are much less expensive than custom developed code, they are supported by the vendor, and are available off the shelf with no waiting for the development to progress. As with any other software purchase, check references, satisfy yourself that the vendor is reputable, verify that the function purchased is what you need, and protect yourself with access to source code.

Although many packaged enhancements are completely outside of the base package, that is, they do not include any modifications to the primary package programs, there are some enhancements that do touch the base code. Be careful here, as the line between enhancement and modification might become a bit blurred. When the change affects the basic package, be especially careful. Assure yourself that the modification to the code was done carefully, is fully documented and supported, and was designed for minimum interference with the base function.

Support is critical. Most software vendors provide fixes and updates to registered users on a regular basis. Any time there is a change made to the vendors' programs, the update process becomes problematic. You bought the basic package from vendor A. You bought an enhancement from vendor B that modifies several of vendor A's programs. Vendor A sends you an update. You must now contact vendor B to find out if its enhancement is impacted by vendor A's update. If so, you must wait until vendor B sends you *its* update before installing vendor A's update. Install A's update first, then B's update, and then test to be sure that both vendors did things right. If there is a problem, call vendor B first; try to eliminate it as the cause of the problem before calling vendor A. All of this obviously takes time and effort.

If you write your own enhancements, the situation is similar, but in this case, vendor B is you. When you receive an update from the primary vendor, you must check the function of your enhancement with the new code, make any necessary changes, and delay the implementation of the update until you have made and fully tested the necessary changes to your add-on.

Of course, you always have the option to forego the updates offered by the vendor, remaining at a particular level of the programs without installing additional updates. This state is known as "down-level" and practically eliminates any possibility of vendor support if you have any problems. This may be a viable option in some circumstances, but most companies, sooner or later, regret a decision to stay down-level. Eventually, your business will change, your market will take a sharp right turn, or new management will want to take advantage of changes in technology, and you will be faced with a difficult decision and an arduous path to migrate to a new solution or the current version of your chosen package.

Modifying Packaged Software

It is perhaps unfortunate but it is true that many packaged software users choose to modify their packages. Why? Many times it is the result of a firm conviction (belief), right or wrong, that their business is unique enough that the standard functions are inadequate. Whether justified or not, the users have demanded changes or additions and the MIS staff (or senior management) has opted to accommodate the requests.

The support and upgrade considerations for modified software are the same as with enhancements but to a greater degree. A modification as I have defined it is a change to the logic of the vendor's programs and thus will be impacted by any update that touches those same programs, and likely will have impacts beyond the individual changed programs.

MRP II systems are, by their nature, integrated. Data are shared. Functions are interdependent. Changes in one area almost always affect one or more other areas within the system. When designing and implementing changes, it is extremely important to thoroughly test for these interactions and understand how data flow within the system.

It is for this reason as well as the others discussed throughout this chapter that modifications tend to be more expensive than anticipated and more difficult to install and support.

Controlling the Change Process

Too often, the MIS department is given the responsibility for deciding what modifications are to be made and which requests are rejected. This can put MIS in a very difficult situation. MIS, being a "customer"-oriented service organization will want to be responsive to its customer—the user. If they refuse to make a requested modification, they can be viewed as unresponsive or uncooperative.

To make matters worse, MIS often reports to the head of another functional department such as finance or administration. If there are thirty requests pending and only enough resource to satisfy twenty, which department do you suppose will get more approvals?

It's really a no-win situation for MIS in such circumstances. They might fully understand the need to limit modifications but there is considerable pressure within the organizational structure to modify. Their measurements are probably tied to the number of projects completed or the amount of code written. To refuse or limit modifications would work against their performance measurement.

My recommendation is to take the basic decisions out of the hands of MIS and put them into a high-level committee with representatives of all major areas of the company. Such a review committee would not be unduly influenced by any particular department and could look more objectively at user requests for modification.

The change control process should be formalized, with every request accompanied by a justification. MIS would obviously be involved in helping the requesting users identify the extent of the change requested and the cost, whether performed in-house, by a contractor, or satisfied by a packaged enhancement. Cost figures should include long-term support considerations. The user also would be required to specify the benefits to be achieved as a result of the change. If the benefits don't outweigh the costs, the committee wouldn't even have to see the request—it would self-destruct in the justification process.

Those requests that make it to the review committee would then be carefully considered, accepted or rejected, and the accepted requests prioritized and scheduled. As a result, MIS has only clearly defined, justified, and preapproved projects and also has the benefits of an unbiased oversight group to resolve conflicts and take the direct pressure of user demands.

Minimizing Impact

There are several significant advantages to keeping your package as "vanilla" as possible. Although some vendors will attempt to support a user who has modified his package, others will simply refuse.

Even if the vendor is willing to try, it is quite difficult to support code that is a combination of standard product and changes instituted by the individual user company. Let's say that your work order release subsystem fails when attempting to initiate an order for an unusually large quantity. If you have made changes to any of the programs involved in the release process, you will not know, at first, whether the failure is a result of your modification or a failure of the vendor's product.

If you call the vendor's support line, one of the first questions likely to be asked is whether there are modifications present. If the answer is yes, and the vendor is still willing to help, you (someone) must identify whether the error occurs in unmodified code. Either the user company or the vendor will try to duplicate the conditions in which the error occurred, not an easy task, and isolate the cause.

If the fault is the vendor's, after it corrects the error, it will be necessary for you to reapply your modifications after the corrected programs are provided by the vendor. If the fault is yours, you will be embarrassed and the vendor will be annoyed, at the very least—and the burden of correction is all yours.

The best modification is the least modification. When you are aware of the consequences of code changes, you will work to minimize the changes to the vendor's code. There are a number of techniques that can be used to achieve this goal. Often, several of the following techniques can be used in combination.

Use separate programs and subroutines for new logic. Insert only a line or two in the vendor's program to "exit" to your function and return when completed.

Copy the vendor's program, insert the change, and retain the original program. This allows you to revert to the original for testing and problem determination. Some systems use a "library list" organization that allows you to have multiple versions of a program on the system. The modified version is put into a list with higher priority than the original version. In this way, the vendor's code remains "pure" in the vendor's library and all modified programs are together in a higher priority "mod" library where they can be easily managed.

Fully document all changes. Documentation cannot be overemphasized and is usually missing, incomplete, or poorly managed. Be sure that the effort required for documentation is included in the costs used in the justification and be sure that it is done. I suggest, at a minimum, internal documentation (program comments), lists of modified programs with description of the changes, annotated program listings with the change highlighted, full documentation of any database changes, updated user documentation (user manuals, procedures), and a log of support issues and maintenance activities.

When removing a line of vendor's code, do not delete it. Inactivate it by changing it to a comment. Every programmer knows how to "comment out" a line of code.

Avoid changes to the vendor's database. Even with relational database management systems, keep changes outside of the vendor's architecture as much as possible. Add a new file or reuse an unused field rather than add a new field to an existing file.

Summary

Even with the best advice, a top-notch systems staff, and a well-executed change management process, the best strategy is to pick a package that provides most of the function you need, adapt your procedures as much as possible in the areas where there are differences, and minimize any changes to the vendor's code if you must modify. The main reasons for buying a package are to be able to take advantage of superior function faster and at less cost than custom programs, to benefit from support services that the vendor and third parties can provide, and to be able to keep up with the technology through the update process. All of these benefits are jeopardized by modifications.

Remember that no package is a perfect fit. Most companies find that careful selection will result in about 80% of the requirements being well satisfied by packaged functions. One key to success is how you choose to handle the other 20%. The 80 : 20 rule works both in your favor and against you. The package can provide 80% of the function (or more) at 20% of the cost (or less), but modifications can supply only a small fraction of the function at a large percentage of the cost.

I have seen many cases of a one hundred thousand dollar package with a half-million dollars in modifications (and equivalent spending in other price ranges). And that is to say nothing of the implementation delays caused by modification, the inability to use standard education and documentation, and the support costs and difficulties.

There is a bright spot on the horizon with new developments in the way packaged software is developed and sold. These coming changes are discussed in Appendix B.

3. Executive Commitment

The large conference room was abuzz. Sixty midlevel managers, representing key departments from the company's seven North American plants were gathering around the long, narrow, cloth-covered tables, sipping coffee and exchanging pleasantries. At precisely eight o'clock, the project manager stepped to the podium and called the meeting to order.

"Gentlemen . . . and ladies, please take a seat so we can get started. We have a full day ahead of us."

The managers did as they were told and soon the room was bathed in an anticipatory hush.

"As you all know, we're here for the next few days to learn everything there is to know about MRP." As expected, this remark was greeted with widespread chuckles and groans.

"Seriously, this is the kickoff for the companywide MRP implementation project and we—all of us together—are the ones that have to make it happen. We have two very experienced and knowledgeable consultants here this week who will educate us for the next three days. Then we'll spend some time working on our project plans for each plant. First on the agenda this morning is a message from our President, Jim MacDonald."

At this point, a large-screen TV was rolled center-stage and a rousing thirty-minute video tape was started. The opening shot was a wide-angle view of the executive's office . . . oversized antique oak desk, burgundy carpet woven with the company logo, plush drapes, and a very presidential looking gentleman sitting behind the desk looking wise and sincere.

The camera slowly zoomed in to a head-and-shoulders shot and the company president started his speech. "This project is the most important effort undertaken by this corporation in recent memory," the gathering was told. The speech went on to say: Competition is tough. The company cannot afford to be left behind. The old corporate systems are not up to the challenge. The new system will cost much less and deliver more. Access to better and more timely information is essential to the company's future. If this effort fails, the company's very survival is at stake.

I was standing at the back of the room, one of the two consultants ready to conduct the class, and I was impressed. The president's message was truly inspirational. The room contained the key managers from the major departments of each plant, gathered as a team to plan and initiate the effort. We were there to start the education process. All of the pieces were in place. This project couldn't fail after such an auspicious start.

After the initial round of classes, I was not involved in the implementation and thus I lost track of this company's project. But that image stayed alive in my memory. As I worked with other companies, I told them about the great start this company had had and about their prospects for success.

Two years later, I happened to bump into the project leader as we were both between flights at O'Hare Airport in Chicago. Over an overpriced beer, I was shocked to learn that the project had stalled at about the nine-month point, had fallen into complete disarray, and the company was floundering badly. The new system was not implemented. The company had continued to patch the old systems to keep it afloat despite the loss of the best people from the data processing staff. This project manager was, in fact, on his way back from a job interview.

What happened? Here was a company with everything going for it. All of the pieces were in place and there was certainly strong motivation in the "do or die" message from the president. What went wrong?

The project leader explained that the president's video message was the beginning and end of his involvement with the project. He apparently felt that he had done his duty and he could move on to other pursuits, which he did.

As the project started moving along, there came a time, as there always will, when there were conflicts, difficulties, and disagreements. Without strong leadership and pressure to continue from top management, the conflicts weren't resolved, the bickering continued, and progress ground to a halt. Meanwhile, the systems staff had received a clear signal that their days were numbered and the best of them had found other jobs. As a result, the company was much worse off than before; the old systems were not being maintained properly because key systems knowledge had departed while the new system was far from the point where it could have provided replacement functionality.

It was clear to all, although few would voice this opinion, that the CEO did not believe what he had said in his video speech. If he had truly believed that this implementation project was that important, he would have stayed involved or at least monitored progress. His behavior did not back up his rhetoric. Absence of strong leadership had allowed the project to stall. Once stalled, it is extremely difficult to reestablish momentum.

Enthusiasm and commitment to a project doesn't "trickle up" or even "bubble up" through an organization. If anything, it trickles down. Although it is certainly important to have commitment throughout the organization, it is the person or persons at the top who set the tone, have enough power to enforce decisions and allocate resources, and from whom all others take their direction, whether explicit or subtle.

This top person whose involvement is so critical is either the Chief Operating Officer (COO) of the company or of similar rank or position. In smaller companies, the COO might also be the Chief Executive Officer (CEO) or President; in a larger company, we may be talking about the plant manager, facility general manager, director of manufacturing, or division manager.

It is not always this top person who is the "champion" of the project. He or she may not have had the initial inspiration or put together the first proposal. But this person must be sold on the idea and be willing to step up and take a very visible position as a firm supporter.

There's an old consultant's story about what the difference is between being committed and being involved. It (the story) is about a pig and a hen strolling down the street together. They happen to notice that there is a "help wanted" sign in the window of a restaurant that is also advertising a bacon-and-egg breakfast special. The hen asks the pig if he's interested in applying for the job. The pig replies "for you, that would be involvement, but for me, it's commitment."

Whichever term you use, the executive that is "pushing" the project must make it clear to everyone in the company that he or she firmly believes that the implementation is critical and that it *will* be successful, no matter what. There will come a time (or more likely several) when the implementation project is in conflict with some other company or departmental goals. This is not to say that the goals of the other project(s) will conflict with those of the implementation effort but that the two objectives will require some of the same resources or that the priorities may get confused. It is also likely that sometime during the course of a lengthy implementation process, enthusiasm will drain away and there is no one else positioned to keep the fires burning than the "top guy."

Having progress and results reported on a regular basis to the executive in charge is also likely to keep people focused on keeping the project moving and meeting objectives so that they won't have to report bad news (more about project management later). Be sure, however, that the reports are meaningful and honest. It does tremendous harm to report false progress to the executive. Sooner or later the truth will come out and later is much worse—the sooner a problem is identified, the easier it will be to solve.

The top executive is seldom an "active" participant in the project itself in the sense of having defined tasks to accomplish and often the executive is not a regular attendee at project meetings. But don't take that as the difference between commitment and involvement. Being committed means having a stake in the game, regardless of the extent of the everyday involvement in the project. If the executive is committed, he or she will want to keep up with progress. He or she will want everyone else in the company to know that he or she is behind the project and is interested in the progress being made.

In 1984, I visited a company in Baltimore, let's call it Klingon Kitchens, that had been working on its MRP II implementation for five years. In this time, despite many hundreds of hours of meetings and work sessions, it was still trying to decide how to load its product definitions (bills-of-material) into the system's database. Now, the way the product configuration is described in the system is certainly an area of concern and can be a critical factor in success or failure of the system, but these people had done nothing but discuss what approach to use for five years. There was obviously something wrong here. These were intelligent, dedicated people who just couldn't seem to get the job done. The reason I was there was because management had scheduled a training session to try to get things moving.

I spent two days with this group, explaining how the system worked, what the bill-of-material entries should look like, and what options were available to handle special cases. I got a plant tour and got to see the actual products—commercial stoves, deep fryers, and the like—and it seemed that there shouldn't be a tremendous problem defining the products in the system.

Six months later there was still no measurable progress.

It turned out that this plant was one of two manufacturing sites within the company that were sharing a single midsized computer. The computer was located in Baltimore and the other site, which was also corporate headquarters, operated on a remote basis from St. Louis. The St. Louis plant had started its implementation project at about the same time as Baltimore but it had been "up and running" for several years. This seemed strange, indeed. St. Louis manufactured a similar product line, actually a little more complicated than Baltimore's. The plant that was remote from the system and therefore handicapped by communications line speed limitations and unreliability, that in fact had built-in excuses for delaying implementation if it wished to use them, was actually a lot more successful than the plant with the advantages.

Why was Baltimore still dithering while St. Louis thrived? The answer is in the location of corporate headquarters. It turns out that the decision to get the system in the first place was made in (surprise!) St. Louis. Baltimore was not involved in the selection and fell under the "not invented here" syndrome. There was absolutely no "ownership" interest at Baltimore and therefore no motivation to succeed. The executive sponsor and primary project booster was also in St. Louis, far enough away (geographically and organizationally) to be ineffective in providing the leadership that this project needed. No amount of training and commitment on the part of the engineers and production people was enough to overcome the Baltimore plant manager's indifference, if not hostility, to the project.

The Executive's Role

The simple answer to the question "What is the executive's role in the implementation?" is "to provide leadership." There are several aspects to providing leadership, however, and there are also aspects to preparing for this role.

To begin, the executive must understand what the project is about, what benefits may be gained, and what it takes to get them. This usually requires some education in the specific kind of system being proposed. A competent executive will certainly know all about how to run his particular kind of business, but it is not a good assumption that the executive will fully understand the issues surrounding the implementation of a system, no matter how smart, experienced, or knowledgeable he or she may be.

Education for the executive may include classes that are specific to the brand of software being installed or generic (not specific to a particular brand of software) classes that cover the concepts, requirements, and impact of the *type* of system. In manufacturing, there are a number of sources for generic MRP II education that do not include information that is specific to any particular brand of software package. In addition to manufacturing systems consultants and professional societies, there are also classes available from colleges and community colleges, and many books have been written on the subject.

Generic concept education not only prepares the executive for his or her role in the project but also helps to set expectations at the right level. A later chapter will focus entirely on expectations, but it is particularly important for the executive to understand what the system can be expected to do and how much effort (and time) it will take to implement it.

Leadership

The executive's leadership responsibilities begin with providing vision and the setting of goals. The best way to begin an implementation project is with a clear vision of just what you expect to achieve. This vision is necessarily high-level in the early stages of the project, but it is important that it is realistic and specific to business benefits.

I have seen many project vision statements that list such goals as "improve the quality of life," "encourage individuality," "respect for the individual," "create a partnership between labor and management," "produce the best widgets in the world," and the like. Although these may be admirable goals, they cannot be achieved simply by implementing a new information system and it is also very difficult to measure progress toward achievement of such "soft" objectives.

Company vision statements usually include these kinds of global objectives and most companies today have a published vision statement to set the tone for the organization. Vision in the project sense is more specific. The company is about to commit a sum of money and substantial resources to the implementation of a system. It is important that the executive leader has a high-level expectation of how the system will affect the company and that he or she publicizes this vision so that all concerned can share in this vision.

More than anything else, the vision statement or high-level goals help to keep things in perspective. They provide a foundation upon which more specific goals can be developed and they provide a framework within which judgments can be made. As the project is more fully defined, there will be items on the list that may fall outside of the original scope of the project. Each specific goal should be viewed in light of whether it supports the overall project objectives. If it doesn't, then it's probably just a distraction. If it is outside of the scope but is really important, either the scope should be changed (dangerous) or another project initiated to address this goal (better). The same is true about conflicting project tasks and other trade-offs. If there is conflict, decisions can be made by comparing the conflicting items to the overall goals of the project and prioritizing or eliminating according to how well each supports these goals.

Finally, the high-level objectives are the starting point for developing more detailed goals. Much of the project planning and management task is a series of refinements—from overall objectives to specific goals, from goals to tasks, from tasks to schedules and detailed activities. Each layer of increasing detail can be validated by how well it supports the layer(s) above.

Project Objectives

Valid project objectives must be of the "hard" variety—things that can be measured—and they must be bottom-line business objectives. Things like reducing inventory, reducing lead time, and improving customer service (specify how this is measured) make great goals if they are appropriate (in context), quantitative, and especially if they are specified as ratios over time.

Let's say a company currently has an annual sales volume of fifty million dollars and a total inventory of ten million dollars. At first glance, you might think a suitable goal would be to reduce inventory by 10%. That can be done easily, and without even installing a computer ... just reduce purchases and cut back on production. Inventory will drop. The trouble with that scenario is that customer service will suffer and efficiency will drop as shortages break out all over the plant (raw materials and finished goods).

Let's refine that goal a little. First, separate the types of inventory: raw materials and components, work-in-process (WIP), and finished goods. Then let's set goals for each group individually based on how well controlled each may be before the changes. Maybe we can target a 20% reduction in parts and materials through the implementation of Material Requirements Planning (MRP), 5% reduction in WIP through better shop-floor control and reduced manufacturing lead time, and 10% reduction in finished goods while increasing customer service to 90%+ (typical results of gaining control of the forecast and master schedule). Tying inventory reduction to a customer service objective ensures that an improvement in one area is not gained at the expense of another important measurement.

This is a much better set of goals in that it is more meaningful and ties the goals to tasks within the project, but we're not quite finished yet. What if the company's sales are increasing at the rate of 15% per year? Would it be fair to expect reductions in the absolute amount of inventory despite the increasing volume? Wouldn't it, in fact, be a major accomplishment to maintain the current inventory level while increasing volume? Actually, to hold inventory at $10 million while increasing sales from $50 million to $57.5 million (a 15% increase) would actually be a *relative* inventory reduction of 13% ($10 million/$50 million = 20% inventory-to-sales ratio; $10 million/$57.5 million = 17.4% ratio; 17.4% is a 13% improvement over 20%).

Now, when would you like to see this improvement? It won't happen overnight. Be realistic in setting goals so that they will motivate and not frustrate. With MRP, for example, it is typical to see an *increase* in inventory shortly after implementation (until you get things under control) and the improvement, when it comes, continues beyond the initial project if the disciplines are maintained. For materials and components, then, a reasonable goals statement might be to reduce the current inventory-to-sales volume ratio by 5% after one year (MRP won't be installed until some months into the project after the database is built and inventory records are in control, then it takes a few months to gain control of MRP), then 10% at eighteen months, 15% at the two-year point, and 20% at thirty months. These numbers will vary based on how out of control things are at the start, how aggressive the project is, and characteristics of the products and processes.

Finally, make sure that the objectives can be achieved as a result of the implementation. Remember that you are implementing an information system that, in and of itself, will provide only marginal benefits, primarily in the areas of reduced clerical effort. The real benefits are the result of better management decisions supported by the availability of information, and improved control and discipline that flow from the implementation process itself. I used to work with a guy whose favorite saying was "If you have a 27% accurate manual inventory system and install a computer, you will end up with a 27% accurate computerized inventory system." A system will not change anything except how the information is handled. The biggest benefits come from using the implementation as a framework to tighten up procedures and develop better communications and cooperation between employees.

In summary, then, the goals should be

- bottom-line business benefits
- specific (finished goods, components materials, and WIP, not just "inventory")
- easily measurable
- ratios where possible
- specified in increments over time
- attributable to the system implementation

High-level goals must come from high-level management. The initial statement that kicks off the project might not be this specific, but whatever vision is provided by the executive(s) should lead, after refinement, to the kind of goals just described. At first, the statements might be more global: "We will implement a new system to provide management information that will support the company's objectives of reducing finished goods inventory while increasing customer service, avoid material shortages while reducing raw material inventory, and shorten manufacturing lead time that will lead to reduced WIP inventory.

Once the goal (vision) is established, it must be communicated throughout the organization. This is pure leadership. The executive tells everyone what this project will do for the company, puts his or her personal imprimatur behind it, and encourages all to cooperate to get it accomplished. The executive must understand, and communicate to the troops, that there will be some inconvenience and some extra work involved in the implementation. Despite the widespread belief in the concept of a "turn-key" system, there is no such thing. No matter how much of the effort is contracted out or included with the package, it is the changes in procedures that allow the users to benefit from the system. Management's acknowledgment of this fact, that there is work and disruption involved, is essential to preparing the employees for the upcoming effort.

This communication of vision and encouragement is not a one-time event. Throughout the life of the implementation project, the executive should stay visibly involved and interested, focused on the importance of the implementation to the survival and profitability of the company.

Another task for the executive is to watch for and manage changes to the organization that will be required to fully implement the new system. Make no mistake, there will be changes required. One company that I worked with a few years ago had a sign prominently displayed in one of the departments that read "If you continue to do things the way you've always done them, you will get the same results you've always gotten." As I said about the 27% accurate inventory system, unless things change, things (results) won't change. Installing a system in and of itself will not drive performance, but changing procedures, processes, and the organization to exploit the new facilities will.

People are naturally resistant to change. No matter how "bad" things are now, they are at least familiar and the introduction of something new is scary. This topic is addressed in more detail later, but the executive should be prepared to support the necessary changes and encourage people throughout the organization to make the commitment and adopt the new ways of doing things.

Changes may also include the structure of the organization itself. As new lines of communication open up, as they will with an integrated system, there must be a willingness to adapt the organizational structure accordingly. People may move from one department to another, the way groups deal with each other will change, and all will be asked to share information more openly than before. All should be more willing to do these things if the "big guy" is openly behind the project.

Executive involvement also supports the allocation of sufficient resources to do the job right. This includes not only the money to buy the hardware and software, but also the time and manpower required to implement the system, and time and money to educate and train the future users. Company executives must realize that a new system will not produce results overnight. Sufficient time must be allocated to put the pieces in place and let the changes become part of the process. Among the most difficult resource allocation decisions are the selection of the project leader and the members of the project team. A good rule of thumb is that the leader and the team members should be the best available—those people that you can least afford to spare from their regular duties.

The executive must also stay informed about project progress. This does not necessarily mean that he or she must attend every project team meeting, but should attend project reviews on a regular basis, receive and review all project status documents, and regularly provide feedback to the team.

An important factor in acceptance of a new system and the willingness to support change is fear of technology, which manifests itself in several ways. First, many people have had limited exposure to computers and are afraid that they will not be able to do the job in a computerized environment. I can see major differences from one generation to another. My father (now retired) had no contact with computers during his working days and won't have anything to do with them. He even has me reset the digital clock in his car in April and October when the time changes. My thirteen-year-old daughter, by contrast, is at this moment using a graphic program to draw Wilma Flintstone on the PC she shares with her older sister.

A consultant friend tells of a situation in which a key production supervisor was the major foot dragger in a system implementation project, stubbornly refusing to accept the system and learn what it could do. When no amount of encouragement, threats, or arguments could move this gentlemen from his reticence, senior company management felt that there were limited options left to them. My consultant friend got the opportunity to talk to the man, away from the plant and confidentially, and learned of the man's fears. More than anything else, the man was afraid that he would "appear stupid" and lose the respect of all the young kids he had to supervise. Some private lessons in basic keyboard techniques and use of the system resolved the problem, along with assurances that his experience and judgment were still valued and respected, even if he was not the fastest typist in town.

The other fear of technology is as a threat to job security. Many of us can remember when computer installation projects were accompanied by the arrival of "efficiency experts" and mass firings. Today, staff reductions are called "downsizing" or the more politically correct "rightsizing," but the effect is the same—more computers equals fewer employees. Whether head-count reduction is a part of the plan or not, there will be a perception among company employees that people will lose their jobs because of the new computer. This perception will discourage open, active participation in the implementation project. It is only natural to assume that a person won't enthusiastically support something that may cost his or her job.

I recommend that the executive behind the project make a clear, published statement that no jobs will be lost as a result of the computer implementation. I know that I'm about to lose a few of you at this point, but I hope you'll bear with me just a little longer.

First, if you justified the system based on a reduction in head count, you may be in serious trouble already. I have seen many projects with a large portion of their justification based on head-count reduction and I have seen just about all of them fail to achieve the reductions expected. It just doesn't happen. The truth of the matter is that many times a successful system implementation will not reduce head count or may, in fact, increase it.

Systems will (or should) reduce the amount of time spent handling paper, filing, retrieving information, and doing other mundane tasks. On the other hand, additional effort is required to enter and maintain data in the system, perform system operation and maintenance tasks, and use the new information that will become available. Often, there is no net reduction in time for these administrative tasks, but the nature of the effort changes. Instead of routine filing and handling paper, people can spend the same time using information for the benefit of the company.

Think about a buyer in the purchasing department. Without system support, information is recorded on cards, forms, or scratch paper, filed by vendor or purchased item, retrieved and sorted by hand, and retyped or copied many times during the course of its use. With a system, vendor information, price quotes, etc., are recorded once and made available on computer screens (throughout the company) when and where needed. Purchase orders are printed out rather than manually typed. Filing is greatly reduced. This does not mean that fewer buyers are needed, however. The buyer's time is redirected from mundane, nonvalue activities and made available for identifying new sources, managing vendor performance, spending more time analyzing needs and purchasing policies, and other value-adding tasks. Although there is no direct cost savings (head-count reduction), there is considerable added value that should show up on the bottom line as more effective purchasing, including higher quality, fewer rejects and disputes, higher availability (fewer late deliveries), and maybe even lower purchase costs.

Back to my main point—it is strongly recommended to *not* justify the system on reduced head count. Not only are you not likely to achieve the head-count reductions, but the employees will find out (they always do) that the system will eliminate jobs and there will be resistance. Why should a worker go to training, make the extra efforts required to load data and establish new procedures, and willingly suffer the trauma and rigors of conversion if the end result is his or her own unemployment.

I recommend, therefore, a clear statement (from the top executive) that the system will not be the cause of lost employment. Some jobs will disappear, however. Your company will need fewer administrative clerks, typists, expediters, and filing clerks. But new positions will be created: material planners, master scheduler, database administrator, cycle counters, and more. Obviously, there will be a need for retraining to turn a filing clerk into a cycle counter or inventory control specialist or to turn a buyer into a material planner or buyer/planner. Be prepared for this and factor this into the project plan (more on this later).

A compensating benefit (in lieu of labor cost reduction) is cost avoidance—the ability to grow (increase business volume) without adding staff or equipment. In addition, there are the other benefits of getting more from the people you already have on staff as in the buyer example.

If you anticipate or experience staff reductions due to business conditions, be sure that it is understood that it is not the system that caused the reductions. This will test your credibility. Be open and honest, and be sure to distance any business-related layoffs from the changes brought about by the system implementation.

I recently conducted a seminar in Missouri at which I encountered Jack Stack, President of Springfield Remanufacturing Company and author of *The Great Game of Business* (Doubleday, New York, 1992). Jack's philosophy is to "open the books" to the employees, teach them how to read financial statements, and make each manager aware of and responsible for the bottom-line impact of his or her contributions to the company's financial health. I also observed, on a study tour of Japanese manufacturers a few years ago, that a key element of the fabled Japanese teamwork is a shared vision based on keeping the employees informed on the state of the business and encouraging their participation through suggestions, empowered teams, and strong ties between profit and compensation (on a company-performance basis). I believe that more openness, particularly in the sharing of business status (financial) information with employees, can be an effective tool in gaining support and participation. It is also necessary to be able to relate the employees' responsibilities and the tasks you are asking them to perform (such as system implementation tasks) to the health of the business and therefore the employees' own compensation and future employment prospects.

Finally, the executive is the final authority for conflict resolution. Hopefully, all disagreements can be settled at lower levels, but there must a published hierarchy for conflict resolution that ends at the sponsoring executive. By establishing this mechanism in advance, not only is conflict resolved in an orderly fashion, but it can actually be avoided. Project participants know that conflicts *will* be resolved and they know that this top executive will get involved if necessary, so there is pressure to cooperate and settle differences before they escalate to this level.

Duties of the Executive

- **Learn**/understand what the system will do (and won't do) and what it takes to achieve the results expected.

- **Provide vision** and **leadership**.
- **Assign** the best project leader.
- **Select**/charge the team.
- **Support the project** (budget sufficient money and resources).
- **Stay involved** through the life of the project.
- **Resolve conflicts** that cannot be settled at lower levels.

4. The Project Team

Sally always sits in the corner at the project team meetings and never speaks unless asked a direct question. Bill, on the other hand, hardly ever shuts up, going on at length about how much he has done for the team and for the company. Louis frowns a lot and usually finds something to complain about, especially when anyone asks his opinion or asks him to take responsibility for a task. Sue is the "recording secretary," dutifully takes notes, and never disagrees with anything. Henry questions every decision and is a master at poking holes in every idea or suggestion. George, the team leader, tries to keep everyone pointed in the same direction, but usually leaves each team meeting with a splitting headache.

Is this project team really a team? Or is it just a group of people that have been "volunteered" to work on the same project? I'd say that it's definitely the latter.

One of the principal impacts of the implementation of an integrated information system, particularly an MRP II system, is that it changes the dynamics of the organization. The integrated nature of the software design forces the disparate segments of the organization to cooperate and to share information—whether they want to or not. If they are unwilling, the implementation will suffer.

We tend to build walls. Each functional department within a company develops an identity and also a sense of individualistic pride, defensiveness, and sometimes paranoia. One reason may be the way we force departments to compete for resources....

It's December second and the final budget meeting has just adjourned. Frank, head of production, emerges from the conference room with a satisfied smile. He has just gotten approval for a CNC[1] *machine and two more production employees for the coming year. Next out of the room is Fred from engineering. His scowl clearly indicates that his request for two new engineering workstations has been turned down. Sally from materials is next. Her neutral expression belies the fact that she, too, is bitterly disappointed. It looks like they'll have to try to find some way to make that old forklift last another year.*

And so it goes. Every year there are winners and losers. Departments are forced to compete against each other for scarce resources and an attitude of "us against them" is bound to result.

Now it's time to install the new MRP system. Anna, the IS manager, calls a meeting and gives a rousing kickoff speech. Because the "big guy" is sitting in the back of the room, all of the department managers agree enthusiastically to participate in the project. When it comes time to do so, however, it may be a different story:

"Engineering will be responsible for entering bills-of-material and routings," Anna announces, "and materials will take charge of item definitions and inventory records."

"My engineers aren't gonna be data entry clerks," huffs Fred, head of engineering. "Either we hire two clerks for the department or you (data processing) will have to key it in for us."

The discussion quickly deteriorates into series of "You had better ..." and "Over my dead body ..." arguments with each manager trying to assert authority. This is obviously more of a mob than a team.

The Team

An integrated system cannot be successfully implemented without a cross-functional team. Coordination and communications are primary aspects of integrated information systems, and only a joint

---

[1] Computerized Numerical Control—a computer-controlled, automated production machine used in manufacturing.

effort of all affected areas of the business can take the responsibility not only for implementing the system, but also for integrating it into their departmental procedures.

The project team should be made up of functional managers from each department that is affected by the implementation project. In the case of MRP II, this includes engineering, production, materials, customer service (sales/marketing and distribution), planning, and finance/administration. These categories will vary with organizational structure (in some companies, materials includes purchasing; in others, production includes materials and purchasing is separate, etc.). Functional managers are department heads.

Depending on the size of the organization, there may actually be a two-tier project team structure: a steering committee of VP-level executives and a working team of department managers. See Figure 4-1. In smaller companies, a single consolidated team is appropriate. Generally, there should be between five and ten members on the team. Larger than that and the team becomes ineffective. Fewer members and there aren't enough skills or decision makers available.

Team members must be decision makers, both in their personalities and in their positions (authority). You don't want a team that can only suggest plans and pass them along for approval. They must be able to make decisions and have the authority to allocate resources and make sure the decisions are carried out. The team members most likely will not "do the work" of implementation. Subteams will be established to address individual project tasks as they are defined and scheduled. More about subteams shortly.

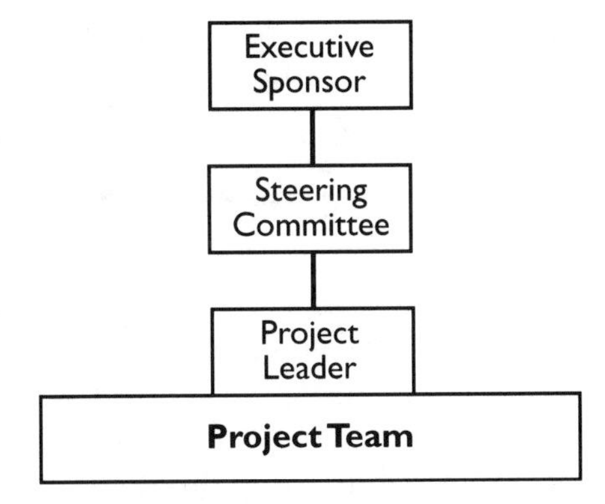

Figure 4-1. High-Level Project Organization

The size and composition of the active team may vary during the life of the project. Although all department heads should remain involved with the team throughout the implementation, there may be times when a manager's department is not yet involved or its portion of the implementation is either completed or between tasks. On these occasions, a team member could revert to a less active role while remaining current on project progress and representing his or her area when appropriate.

Other than the previously mentioned exception, project team meeting attendance should be mandatory. Meetings are typically weekly and are scheduled so that all members know when they will be and can schedule around them. If a member must be away on a scheduled meeting date, either telephone conferencing should be arranged or a suitable substitute can represent the missing member. The substitute must be delegated the proper authority to make decisions and allocate departmental resources as needed to support the project.

Generally speaking, you cannot form a project team by putting a sign-up sheet on the bulletin board. Team members are selected for their skills and because of their positions. Because they are the backbone of the project, however, they should be the most enthusiastic supporters of the implementation. This could be a problem if not all department heads are "sold" on the systems's benefits. Refer back to the previous chapter on executive commitment. If the "big guy" is sold, that makes it a lot easier to get the people who report to him onto the bandwagon.

One of the general rules for a successful team is that the team must have autonomy—the ability to make decisions and carry them out without interference from other management authorities. With the primary managers making up the team and the COO fully committed, this cannot be a problem.

Another teamwork rule warns against domineering individuals. The team must work together and share the load and responsibility. If one team member is much stronger than the others, the team becomes a dictatorship and the motivation for enthusiastic participation goes away. As much as possible, team members should be peers within the overall organization.

What makes the difference between a group and a team? Interdependence and a common goal. The team is assembled for the purpose of implementing a system. The implementation provides the common goal—assuming that all team members want it. The interdependence comes from the realization that it cannot be done by any one individual or even any one department.

Integrated systems generate a synergy in which each part of the organization contributes to the effort and each receives a benefit that exceeds its contribution. The inventory department, for example, reports activities and maintains the warehouse records. As a reward for this effort, the system provides information that is a benefit to the department such as sorted reports, usage statistics, valuations, and so on. The same is true in the production department. It enters activity data and the system returns value in the form of status information, schedules, costs, priorities, and so on. Now, when both inventory and production are tied together in the same system, inventory can have advance warning of production requirements and be made aware of incoming products, and production can check on potential shortages. Each department realizes an additional gain from the participation of the other department(s) via integration.

The team members, then, are tied together by this potential gain. They must coordinate their efforts for their own benefit as well as that of the other team members (departments). Only cooperation can make it happen effectively and each must be convinced that the system will provide significant value.

Assembling a functioning team from key departmental managers is not always easy. Quite often, these people are in an adversarial relationship with each other—competing for resources and accusing each other of not supporting their individual department's needs adequately.

(Production: "We couldn't finish that customer order because materials didn't get the parts in on time." Materials: "Ask Purchasing why the vendor didn't deliver." Purchasing: "It's not our fault; quality hasn't certified the vendor yet." Quality: "We can't certify until engineering gives us the final specifications." Engineering: "Production changed the machining process, so we have to reinspect the materials.")

The thing that will bring the departments together is a realization that the system will become an essential tool for the company to become or remain competitive. The survival and overall health of the company are at stake (and if they aren't, at least arguably, why are you doing this project in the first place?) as well as the promise of improvement in processes, procedures, access to information, or whatever the more direct results of the implementation will be. This is the shared vision.

What Does the Team Do?

The first order of business is for team members to understand what they are about to do. The team must be the first ones educated on the what, how, and why of the system to be implemented, and this training should come very early in the process as the next task is to develop the project plan. The team will probably also need training in the operation of a team. There are courses, books, and videos available that outline how a team should be organized, tips on effective team meetings, and how to manage communication and coordination of team members' activities.

It may sound a little silly, but you should also consider some training in how to conduct a meeting. Most of us spend a good portion of our lives in meetings, and involvement in the project team will add more to the collection. Why not make the meetings as efficient and worthwhile as possible?

The project planning process should have already started in that the initial justification (see the previous chapter) contains the seeds for the overall plan—the high-level goals and objectives. It is the team's responsibility to add detail and to identify the specific steps (tasks) that are to be completed to implement the system and achieve the goals. Once the plan is in place, the project team pulls together the task teams and manages progress.

While detailing the project plan, the team should also put together the education and training plan for the rest of the company as well as any additional education for itself. There is no better time to identify what education and training will be needed than when the detailed plan is being developed.

As mentioned before, the task teams actually "do" the implementation, whereas the project team is the planning and management authority. Task teams coincide with tasks identified on the plan. Each task addresses a specific accomplishment, one that is important to the implementation and one that can be completed in a reasonable time by a manageable group of people. Each task team has a leader who holds the responsibility for completion of the task on time, and each task has a firm completion date assigned. Although the tasks are delegated to subteams with a responsible leader, the project team retains overall responsibility for the project. Each task will fall primarily within the area of one project team member (although many tasks will cross departmental boundaries) who will act as project team coordinator for that task. The entire team, however, should feel responsible for the completion of each task since all team members contribute to the overall success of the project.

Joe, an engineering tech, was called into his manager's office at 3:45 on Monday afternoon, right after the project team meeting.

"Joe, the project team laid out this implementation task and we caught it because it falls into the engineering area. What I need you to do is review the part numbering system and come up with a set of guidelines that will be published and enforced for all new parts from now on. You have two weeks to get it done."

"Will I have any help on this?"

"I'm afraid not. Everyone else is tied up with the new product release. I'll tell the engineers that they should cooperate and if you have any problems let me know."

"How about my regular duties? Can I get some help or relief on those for the next two weeks?"

"No can do. Like I said, this new product roll-out takes precedence. I want you to take care of this in your spare time. It shouldn't take that long. I know I can count on you."

Two weeks later, on Monday morning:

"Listen, this is a much bigger job than you might have thought. I've put in more than thirty hours already and there's still three more product lines to review before I can finalize the numbering system."

"Well, Okay, I'll get you another week, but it had better be finished by then."

"Thanks, boss. I promise you it'll be done next Monday."

There are several major problems here. First, the task was assigned without buy-in by the task leader. Just as dictatorship doesn't work in the project team, it won't work with the tasks either. The task leader should have the opportunity to help define the details of the task and agree to the amount of effort required and over what duration that effort can be applied. Not only will the task leader feel more a part of the project and therefore be more willing to give extra effort, he or she may also be able to offer valuable advice and suggestions as to how to accomplish the task.

One thing that I have noticed in working with many American manufacturing companies is that we tend not to exploit the experience and expertise of line-level employees. I had the opportunity to tour a number of factories in Japan in 1993 and it is striking how effectively this resource (employee knowledge and experience) is utilized there through such things as *Kaisen*, the suggestion-based continuous improvement programs, and Total Productive Maintenance (TPM), which gets direct production workers more involved in design and maintenance of work areas, work environment improvements, and team-based mutual support.

Poor Joe was assigned the task by his boss without Joe really understanding its importance or the size of the job. The way it was presented, he couldn't really refuse, and he didn't understand the task well enough to debate the schedule or resource allocation. If he had been able to help in the planning, he might have been able to identify that it was a fifty-hour job and either lobby for more time or change priorities to be able to get it done by the deadline.

Second, the job was to be done by Joe, alone, in his "spare time." Nobody in any company that I have worked with in the last fifteen years has had any spare time. All companies are running "lean and mean" or they're running themselves out of business. Yet, major system implementation projects are completed by these same people that were already working to the limit of their capabilities. How can this be?

If the job or task is important enough, it will get done. There is a certain elasticity of time and effort that allows people to do the seemingly impossible on a regular basis. Priority has a lot to do with it. The extra fifty hours for this task can come from some additional after-hours effort, from postponing other less-important tasks, by being more efficient on other duties, by eliminating some things that are routinely done but could be skipped sometimes if need be, and from other similar sources. This is the way things get done, and the project team members and other participants will likely use the same methods to create time for the project . . . if it is important enough.

There must be "buy-in" (as there was not in this example) for the person to "make" the time. The best way to get buy-in is to provide an adequate answer to the question: "What's in it for me?" Show the benefits. Help the person see how his or her job will be easier, or how things will be better after the implementation is completed. And, by all means, involve them in the planning.

Another problem in the preceding scenario, and this one is especially serious, is that the problem in completing the task was not identified until the schedule was compromised. As soon as a potential problem is identified, it is important to bring it to the attention of the project team so that it can be resolved and the task completed on time. To find out on the due date that the task isn't done is absolutely unacceptable. Project planning and management will be discussed in more detail in the following chapters, but this is a key point.

Another project team responsibility is to continue to sell the project to the rest of the company. This is done primarily by keeping its own interest and commitment high and making sure that others are aware of it. There is no doubt that this is a selling job—primarily by spreading the word about accomplishments and expected benefits.

Many companies use posters, slogans, buttons, newsletters and/or banners to try to build and maintain project enthusiasm. Although these devices can sometimes be useful, too often banners and other gimmicks are a sign that management has done all they are going to do. Most people have grown cynical about this kind of thing and more often than not the cynicism is justified. I have toured many plants where the rafters are adorned with colorful proclamations of a commitment to quality or excellence or something of the sort, and there is no further evidence in the plant of any such commitment. If your company has any history whatsoever of this sort of thing, I advise you to stay away from the contrivances that were used before.

It is important to disseminate information about project goals and progress, but don't get too carried away with style or appearance—focus on the information itself. Personally, I like large charts, prominently displayed. There should be measurable items in the plan (more on this in the next chapter) that can be charted and percentage completion and achievement of milestones are also good candidates for a big wall chart. Also consider awards and informal get-togethers (picnics, simple parties, luncheons) as mechanisms for keeping progress visible and enthusiasm high.

The other team responsibility is to be a part of the conflict-resolution process. Hopefully, most conflicts can be resolved at the project team level. The team members should have the authority to commit resources and make decisions for their areas of the business, and the shared vision should motivate them to cooperate with the other team members for the common good and the good of the project.

If such resolution is not forthcoming, there are other levels in the conflict-resolution process. The resolution process should be clearly outlined for all to see before the mechanism is needed. The first level of conflict resolution lies within the project team. If the team cannot work out the problem, it is escalated to the executive steering committee (if there is one). After that, the executive involved in the project, typically the COO, is the final authority. Everyone should realize that it is desirable to resolve conflicts at as low a level as possible.

Project Team Meetings

The project team should meet every week without fail. There should be no excuse for canceling a project team meeting. The excuse I hear most often is "There wasn't anything to report." That's the best reason I've ever heard to *have* the meeting—to find out what went wrong. There must be progress to report every week or the schedule is too slack.

Implementing a system is a big job. A comprehensive implementation like MRP II often takes eighteen months or more (details about project planning are in the next chapter). If the implementation schedule is not aggressive, enthusiasm will dwindle. The schedule will be too long to keep the required level of interest. Lack of a sense of urgency will lead to a casual attitude about project tasks and deadlines. As a result, it simply won't happen—that is, the project will stall and never reach completion.

In any case, the project team meeting must be well organized and efficient so as not to waste these valuable people's precious time. Have an agenda (published ahead of time so everyone is prepared). Set a starting and an ending time and stick to them.

The meeting should typically last no more than one to one-and-one-half hours. If special problems must be addressed that won't fit that format, special meetings can be set up for those who are involved and the results or recommendations reported to the team at the weekly meeting. There should be no major surprises that justify an extra, *unscheduled* project team meeting.

The primary business of the team meeting is to review the week's accomplishments, coordinate on upcoming tasks, and address any issues that need team input or resolution. Task team reports are passed through the project team member whose area the task falls into or by the task team leader if appropriate. The project team leader documents project status and passes this information up to company management.

The COO (project executive) is free to attend any project team meeting, but in most case does so only occasionally. He or she should meet with the team leader at least weekly to stay "in the loop" and lend support and encouragement to the project.

Remember that all team members (with the possible exception of the project leader) have "regular" jobs to do, so be sure that team meeting time is well spent.

Team Responsibilities

- Learn/understand
- Develop a project plan
- Develop an education plan
- Assign/monitor task teams
- Manage the implementation
- Keep enthusiasm and interest high
- Resolve conflicts

To summarize the project team's duties, it is the team that makes it happen. After the team understands the scope of the project and enough of the technical details to be able to perform its duties, it takes responsibility for developing the complete project plan (next chapter). This plan will evolve from the high-level statement of goals and objectives that got the project started and it must support those objectives. The project plan includes schedules—when things should happen in relation to the overall project schedule and any interdependencies between tasks or phases of the project. At the same time, appropriate education and training can be identified and a schedule set up for it as well. Finally, task teams can be established for the earlier tasks.

IS and the Project Team

You may have noticed that Information Systems (the folks that manage the computer—referred to as IS here, but could also be called Data Processing, Systems, Information Technology, or any number of other things, some of which are not printable) is not listed as a member of the project team. See Figure 4-2. This is an intentional omission. IS *should* be involved, but as a nonvoting member. Let me explain.

Obviously, a computer is involved in a system implementation project. The hardware and software usually make up the major up-front investment. Because of this, there is a natural tendency to think of this as an IS project and put IS in charge. The problem with that approach is that it discourages ownership or buy-in by the users.

Alice is purchasing manager at Oak Accessories Inc. (OAI) and she, like most other folks, is very busy. Yet, she has been "volunteered" to be the task team leader for building the vendor master file in the new computer system. She has been to a class on the new purchasing software that is a part of the system; she understands how things will be different (perhaps even better) after the implementation project is completed. She is even looking forward to having the new system in place and operating.

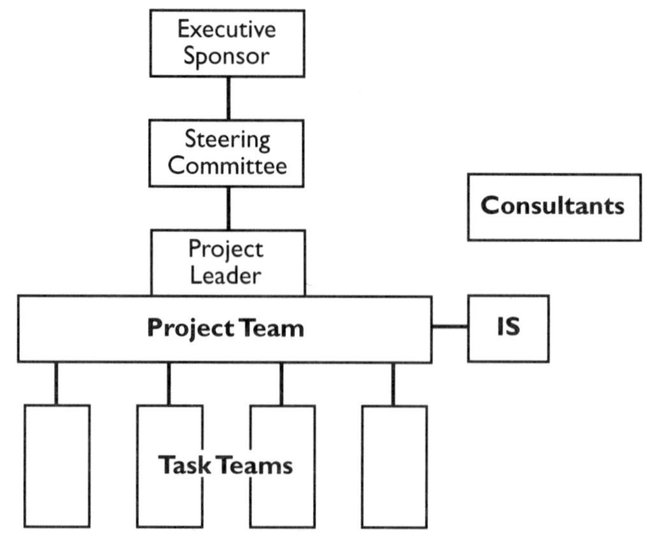

Figure 4-2. Project Organization

The project team is headed by IS (it's its new computer, after all), but she was consulted about the vendor file task before accepting the assignment. She agreed to the level of effort and the schedule.

That was three weeks ago and things have been progressing reasonably well up until today when the "stuff" hit the fan. There was a fire at an important supplier's main warehouse and Alice spent the entire day dealing with the crisis. It's now four-thirty and she is just starting on the day's regular duties. She is already resigned to the fact that she'll be here well into the evening, but even so, not everything will get finished. It's time to prioritize.

"Let's see," she mutters, "there's the month-end reports that can wait another day. There's two dozen POs to get out—those can't wait. I could stay another hour or two after that to work on the vendor file, but that's just to keep IS off my back, so the heck with it. If I let that go, I can be home in time for that movie on cable TV."

Alice doesn't feel a sense of ownership toward the system because it isn't hers. It belongs to IS, and therefore system-related tasks get treated like any other obligation to another department. If she really felt that (at least a part of) the system belonged to her and she was doing the work for her own benefit, she might have felt inclined to skip the movie and work on the vendor file for a while after finishing the batch of POs.

Obviously, IS has an important role to play in the implementation but it must be careful to position itself as a *support* function and not the leader nor even an active participant in the management of the project.

My suggestion that the IS representative be a nonvoting member of the project team is a handy compromise. In this position, IS fully participates in the process without leading or making decisions. It is present to answer technical questions and offer advice to the team relative to system issues. It stays fully informed about progress and system requirements. But its support posture helps keep the functional people in the lead—as they must be.

Consultants/Advisors

There are also no outsiders on the team. Although outside resources may be invited to assist with certain team and project duties, they must also remain in a strictly supporting role. Chapter 13 deals with consultants and other outside support resources.

The Project Leader

Every project must have a leader, someone who has the ultimate responsibility for coordinating and managing the project. It is really *the team* that has the responsibility, as you have just seen, but the team must have a leader or coordinator. The title project team leader could also be used. Project manager, as a title, sends the wrong message.

The project leader's primary tasks are to lead the project team, keep the plan (documentation) up to date, and provide a liaison between the team (the project) and corporate management. Often, the project leader assignment is a full-time job but is usually for a limited duration. This presents a challenge. You want the best individual you can find, but a person that valuable cannot and won't want to be shunted into a dead-end job.

Depending on the size and importance of the project, the leader may be a full-time position. Your initial project justification will tell you whether or not full-time leadership is warranted. If the company is spending in excess of a million dollars and expects a return of hundreds of thousands per year, a full-time position is probably easily justified, at least for the year or two that the implementation project will be in full swing. Whatever the decision, include the cost as part of your justification. For a more modest project, or in a smaller company where full-time project leadership is not affordable, be sure that the assigned leader can devote enough time to do the job right.

The project leader should come from the ranks of present employees, must be a respected and capable individual, and must have good communications skills and be able to work with all levels of people from line workers up to the top company executives. The leader should come from an operational department of the company, not IS.

Some companies look outside to recruit a project leader with system implementation experience. I believe this is a mistake (unless there simply isn't anyone within the company capable of doing the job—a highly unlikely possibility). The system can be learned and there will be resources available for this learning, including the vendor's support staff, classes, system documentation, and consultants. On the other hand, an outsider would have to learn not only the company's products, processes, and ways of doing business, but also the organization, personalities, and politics. These last factors only can be learned from experience in the company environment. There are no classes or documentation available that describes the dynamics of your company and its internal politics.

The primary characteristics of the leader are communications skills and the ability to get things done. Some people are "doers" and some aren't. The project leader must have that undefinable quality that allows him or her to overcome obstacles, sail through adversity, avoid getting frustrated or bogged down, and generally make things happen. Although a strong personality is required, the individual should not be so domineering as to overwhelm the rest of the team. This shouldn't be a problem, because the other team members are also leaders—senior managers—whose demonstrated leadership capabilities helped them attain their current positions.

The project leader communicates both up and down from the team, bringing status information to the executives and coordinating with task teams and the rest of the company. A positive attitude and outgoing personality are assets.

For the same reasons that IS must take a *supporting* role in the project team, the project leader cannot be from the IS department. This is one of the most common mistakes made in organizing an implementation project. Because there is a computer involved, IS naturally assumes the leading role in selection, acquisition, and implementation. The IS area, of necessity, will be actively involved in technical aspects of the project, many of which occur in the early stages. IS will undoubtedly participate in identifying candidate vendors, evaluating system performance factors, verifying proposed hardware and operating system adequacy, and specifying support requirements. IS will also take responsibility for physically installing the system hardware and software or supervising the vendors who do so. Then it will support the operation of the system. The list goes on. Obviously, IS is a key part of the implementation; it just cannot be seen as the driving force behind it. The users must have ownership.

If the team leader has done a good job and brought the project in on time, he or she deserves a reward for a job well done. The problem is that this individual has just worked himself or herself out of a job. The project leader will have been drawn from a responsible position within the company, which position has now been filled by someone else for the year or two that the project was in process. It wouldn't be fair to displace the replacement and put the project leader back into the same position, would it? That's not much of a reward for the efforts and accomplishments of either individual.

At the start of the project, both company management and the project leader should have some idea about what awaits the leader at the end of the project. The experience of leading the team (and the company) through a successful implementation, the high visibility of the leader position, and the opportunity the leader has had in working with all areas of the company should prepare this person for a responsible position in management or the lead position in another important project. Avoid embarrassment or disappointment at the end of the project. Plan for this transition early.

By giving the project leader a reason to want the project finished (successfully and on time), you will be helping to motivate the leader toward success. If there is doubt about what will happen at the end of the project or the personal result is not that attractive, the leader will not have the proper incentive to "work himself or herself out of a job." The leader must be anxious to finish the project and finish it successfully.

James was the project leader for the MRP II system implementation at S. S. Smith Company (SSS), having previously worked his way up from forklift driver to warehouse manager to production scheduler. James knew the company well and understood the politics and personalities. For the first few months of the project, things went quite well; the team pulled together and developed the plan, scheduled lots of training classes, and launched dozens of task teams that accomplished their objectives.

As the project entered its second phase, which addressed the controversial customer order handling and distribution area, the team began to come apart at the seams. Amanda, the team member from customer service, started taking a stronger hand in detailed planning and became intransigent on certain issues. Bill from the warehouse (distribution) disagreed with some of Amanda's decisions and when Amanda wouldn't compromise, started to lobby for support for his agenda among the other team members. Amanda was off on her own quests now, so she did not know of this lobbying nor did she try to build support for her own ideas. When the issues finally came to a head, James was unable to engineer a compromise. By this time, the team had deteriorated into warring factions and any respect or influence James had had was now gone. The project stalled. The company president eventually got involved and took decisions away from the team, making key decisions himself. Nobody was happy with the results.

James failed to detect the growing conflict within the team or, if he did notice, he did nothing to stop it from getting out of hand. He did not use his position and his leadership skills to bring the parties together and resolve the disagreements. He failed to keep the team focused on the successful completion of the project and the latent rivalry and distrust surfaced to the detriment of the project. Just to let you know how this whole sordid affair ended, James was replaced as team leader, and went back to the planning department in a newly created master scheduling position. Within six months, he was gone.

Who should be the project leader? Look around and pick out the person you can least afford to take away from his current duties. Look for the one who always finishes assignments on time, is organized, and is respected by his or her peers, subordinates, and supervisors. Identify the person whom other employees go to for help, advice, and information, no matter what that person's title happens to be. But don't go looking in the IS department.

Ownership

Participants must really *want* to get the system in and working and this only occurs when the benefits are visible and highly desirable.

Benefits are identified in the project planning process, but often they are global, indirect, or just not specific and identifiable enough for the individual employee to get excited about. Project planners have to be aware of the "what's in it for me" factor. Although few people will express this question so directly, all will be subtly influenced by the perceived *personal* value of the project and the system. The best thing to do is to paint a picture of how each user's life will be better after the system implementation is complete. Sometimes this is extremely easy, but other times it can be a real challenge.

Ownership begins with departmental participation in the project planning and management (the project team). The departmental representative speaks for his or her area and makes sure that the solution works for his or her needs. This message is then carried back to the other members of the department and "sold" as a worthwhile effort, worthy of everyone's wholehearted support.

There is an unfortunate "law" that recognizes that the best overall (optimum) solution in a complex environment will not be the optimum for each individual aspect of the complex system. Stated another way, what's best for the company as a whole might not be the absolute best for materials or production or distribution. This is a major reason why a cross-functional team implementation effort is essential. If each department is allowed to implement its own portion of the system in the way that is best for its area, the overall result will not be what's best for the company. The interactions get in the way. One of the primary functions of the team is to recognize these interactions and devise solutions that optimize the *overall* impact of the system.

Team members take this message back to their departments. If there are compromises necessary for the "greater good," the team member must be prepared to explain why things are to be done in this way and convince departmental employees of the value of their agreement and support.

5. Implementation Planning

The consultant's magazine advertisement offered a system review for a modest (fixed) price, including a site visit, interviews with key users, and a written report outlining the current status of the system implementation and recommendations for improvement.

The consultant arrives at the plant and is met in the lobby by Tom, the company Information Systems (IS) manager. Tom arranged for the system review after reading the ad and recommending this consultant and his service to his boss, the plant manager. Tom takes the consultant on a tour of the plant and outlines the situation.

"We selected and installed this packaged MRP II system five years ago, but we're not really using it. I'd like to upgrade the hardware, but the board won't spend any money until we get more use out of what we've already got."

Although the company had purchased twelve application modules, only the basic financial applications and inventory management were in use. Half of the applications purchased, the ones that would provide the biggest benefits, were installed on the first day but have remained untouched ever since.

When asked why, Tom could only shrug. "We had some help from the software vendor when the system was first installed. They converted the data from our old system and ran a series of classes for the users. But things got busy and we just never got around to using those other applications.

"The materials manager keeps telling me that he wants to get MRP working but something always seems to come up that takes precedence."

You would be surprised at how common this situation is. Many companies have spent large sums of money on computer hardware and software but have relatively little return for that investment. In many other cases, companies will buy a dozen software applications but only actually implement a very few of them. In most cases, the primary cause is failure to effectively plan and manage the implementation process.

It has been said that people don't plan to fail; they fail to plan. It is startling how many times a major project, with considerable expenditures and of critical importance, is expected to "just happen" without a detailed plan and an appropriate management structure to shepherd it through to completion. Perhaps part of the problem is the perception that computer systems can "do" things. They cannot. Computers only manage information, or turn data into information so that a user or manager can be better informed and hopefully make better decisions.

In any case, for an implementation project to succeed, the whole project must be planned out and then managed to completion. Most of the better software vendors will include some assistance with the project planning and management along with the purchase of the system (usually called an implementation methodology). Often they will provide sample time-line charts (Gantt charts), critical path diagrams, and/or "templates" to be used with a PC-based project management package. These sample plans are certainly helpful, but such "generic" project plans must be adapted to the specific needs and circumstances of the current situation to be useful.

A major impediment to software vendors providing real help in recognizing the need for training and implementation assistance is the highly competitive nature of the market. In many cases, several vendors are aggressively vying for the business and the decision often comes down to who has the lowest price (I've already made the point that features and function are not significantly different between most packages in a given market). One easy way to lower the initial price of a system is to ignore or underestimate the costs of implementation. You're probably already ahead of me here ... this approach is, of course unrealistic—"penny wise and pound foolish." If the system is not properly implemented, that is, you don't take the time and spend the money to do it right, a much higher price will be paid in lost benefits.

So, given that vendors have incentives to keep implementation costs down for the sake of a more attractive apparent purchase price, the vendors, who are probably in the best position to know what it takes to be successful with their systems, are inhibited from transferring this information to the customer. The customer, consequently, goes into the implementation project with an overly optimistic view of the time and effort required.

By portraying the system as "user-friendly," "easy to use," "intuitive," or "turn-key," the vendors are also setting the customer's expectations to not pay attention to implementation planning, user training, and other important implementation tasks.

Implementation Planning

"Here's the plan," Jim smiled as he placed the document on his boss' desk. "All 500 pages of it."

Jim's boss, Fred, assumed a properly supportive expression. "Well, if we're judging it by the pound, it's certainly a winner. Sit down and tell me about it."

Jim pulled out a diagram of the new system layout that looked like a cross between a Boston road map and a plate of spaghetti. After 45 minutes, Fred held up his hands, palms out, and said, "Whoa, Let's take it up a level and look at the separate applications you want to install and how they relate to each other."

Planning is essential and it begins before the system is purchased. Indeed, the earlier the planning begins, the better. A properly thought-out and structured effort will begin with a set of objectives, some general layout of requirements (what must be done) and schedules, and a general idea of what impact there will be on the organization during and after the changes. Subsequent phases of the process will expand and refine these general ideas into a detailed plan that can be used to allocate resources, organize the effort, and track progress.

The executive sponsor develops the highest-level, first-cut structure that defines the overall objectives (Chapter 3). From here, the project probably isn't justified and funded until a second level of detail is added—sometimes called a "business case." The business case is then used as an outline or road map for the detailed project plan. The project leader and project team develop and maintain the detailed plan. Usually, a relatively small initial team is assembled to put the business case together prior to project approval.

The Business Case

Sometimes called a business impact statement, the business case is a refinement of the initial goal (vision) statement but not as detailed as the completed project plan will be. The primary purpose of the business case is to document the projected costs and the return from the project, in both money and time. It is a formal statement of budget (money and other resources) required and high-level schedule, with enough detail about the expected results to allow the company's management team and board of directors to approve the project and allocate resources to support it.

Beyond budget justification, however, the business case provides the first opportunity to document exactly what is intended and what it's going to take to get it. It is a logical intermediate step between the initial vision (very high level) and the full detail of the project plan.

Overall goals (inventory reduction, improved customer service, reduced lead time) are quantified and time-phased in the business case. They are quantified to the extent that they can be used as measurements of progress and achievement. Beyond that, the business case's documented, measurable benefits become the framework that can be used to keep the project on track.

As the detailed project plan is developed, keep referring back to the business case to see if each task fits within the overall project objectives and time schedule. There will be strong temptations to deviate from the original goals. Some tasks that come up for consideration will appear reasonable and valuable, but, if they are outside of the project framework, beware. They will distract from other tasks that *are* a part of the project, using up resources that were allocated to other jobs. These distractions can lead to schedule slips and cost overruns.

Even though the business case (or perhaps an early, less detailed version of it) is used for justification and budgeting, its more important use is to provide incentive to keep the project going during difficult periods, keep all activities focused on the real project goals, and generally keep things in perspective.

Too often, this step is omitted in the planning process. It may seem redundant, in light of all of the other planning that is recommended herein, but I see it as just one logical step in the process of refining high-level vision into doable tasks. Without a business case, however, it is far too easy to stray away from the initial goals. A year or so after the board agrees to spend X dollars, they probably will want to see just what they got from this investment. It is a great help if both the project team and the board have the same expectations, and an even greater help to the project team to have specific measurable objectives to monitor progress along the way.

Other reasons why the business case is not done include failure to recognize its importance, not knowing what it is or how to do it, and a reluctance to commit to improvement that may not occur. By documenting the expected results of the project, a highly visible commitment is made. If for any reason the goals are not met, the team might be seen to have failed—and nobody wants to fail. If we just don't commit to specific improvements up front, then we can define any results we do achieve as success. This is like shooting an arrow at a blank wall and then drawing a bull's-eye around the point where it lands.

The business case documentation is also an early checkpoint to see if the project goals really support the company's goals and direction. It sounds like a noble objective to increase efficiency and reduce the labor involved in handling inventory (with warehouse automation, for example), but another project or goal might be to reduce inventory to the point that there is very little of it left to be shuttled around in those self-guided robotic vehicles and carousels.

Once again, let me emphasize that project goals must be measurable, in perspective (coordinated with other effects and goals), time-phased (to provide intermediate checkpoints), and directly related to the changes to be made. (See Chapter 3.) For an MRP II project, typical goals (and results) may include the following:

- Inventory reductions of 10% to 50%: The improvement will depend on how bad off things were when you started and how aggressively you pursue procedural discipline. Inventory improvement should be measured by category—raw materials and parts, work-in-process, finished goods—because different tools and processes control each kind of inventory.
- Customer service increases to greater than 90%: This, again, depends on the starting point, goal, and discipline. This goal must be coordinated with inventory targets because they are

directly related. Service always can be improved with an increase in inventory. To increase inventory *and* improve service is the real challenge.

- Lead-time reductions are very important as a key to competitive advantage (responsiveness, flexibility).
- Productivity improvements are often sought, but should be viewed in the context of how important (or unimportant) direct labor is in the overall scheme of things and what the cost is in terms of lost flexibility, increased inventory that may result, and so on.
- Cost reductions are at the bottom of this list intentionally. Virtually all improvements will reduce costs, except those that are aimed at improving the top line (increasing sales). It is much more productive to focus on the management of value-adding activities and the elimination of nonvalue-adding activities than to look at only costs themselves.

The business case looks at both sides of the ledger—estimates of the benefits (results) compared to the cost to get those results. For business case purposes, cost estimates will be based on known values such as computer hardware and software plus a "best guess" at the time and effort required. As the detailed plan is developed later, these estimates will be fleshed-out and validated.

To develop the business case estimates, you will probably require the assistance of experienced implementers. The system vendor is a good source of information, but be sure that the estimates from the supplier are not overly optimistic. The supplier may be tempted to understate implementation costs to make the project appear more affordable. Another source, of course, is an independent system implementation consultant. Be sure the consultant has experience in planning and managing similar implementations in your industry.

The justification included in a business case is most often intended to be conservative. Because the team is committing, on paper, to the expected results, it is normal to add a little "protection" by overestimating costs and underestimating return. It happens, however, that even a conservative estimate fails because of unforeseen problems or forgotten expenses. An experienced consultant can help point out missing items or unrealistic expectations, but remember that you know your company, its business, and the personalities involved much better than an outside advisor can ever know them. Be as realistic as you possibly can and then add your buffers to make the case more conservative. If the result does not justify the expense, and revisiting the numbers yields the same conclusion, then you are much better off knowing that before spending the time, effort, and money.

The Project Plan

The project plan need not be "pretty" or elaborate, but it must be complete, detailed, and realistic. The detail plan should be put together by the project team. Documentation of the plan and updating it to reflect progress are the responsibility of the project team leader. The actual "doing" of the tasks falls to the members of various tasks teams or subteams that will be established as needed, exist only for the duration of the task, and disband when the task is complete.

Let's say that we will be implementing a simple inventory management application. The overall goals of the project are to have accurate inventory information available at all times and conveniently available to all users who need it. The benefits include better customer service because we will be able to accurately tell a customer what we have on hand, reduced cost because we won't "lose" parts and therefore overbuy, and time savings due to not having to search for parts that aren't there.

Although these goals might make the board of directors happy, I'm not sure that the store room supervisor will be. However, I believe this person could be made to envision a time in the future when he would not spend a large portion of his day arguing with others about whether or not there are any widgets in stock. He'd probably also appreciate having balance information at his finger tips, not having to keep up the current card file, getting less hassle from people disputing his numbers, and enjoying the labor savings previously mentioned.

The inventory application won't do any of these things, however. If the current card system is 50% accurate and it is simply converted a computer file, the new system will likely be about 50% accurate. Real improvement comes only when procedural discipline improves. Fortunately, the disciplines required by computerized systems can be the excuse for improving procedures and processes in conjunction with the system implementation project. To help make the system successful (due in large part to the vision of better days ahead), it is often possible to establish new procedures and begin to pay attention to the details that make records accurate. So, under the auspices of implementation, disciplines are introduced and adopted that make the computerized version of the inventory system 95% accurate rather than the 50% that existed under the manual system.

What does this project plan look like? The project team will outline the entire effort and identify manageable subtasks such as building the item master file, loading the initial balances, establishing transaction handling procedures for receiving, issuing parts to production, and filling customer orders, setting up a cycle counting program, and so on. They must then identify dependencies (the item master must be in place before balances can be loaded, for example). The level of effort required for each task is then estimated in labor hours. The amount of time available to apply to the task is factored in to determine the duration (elapsed time) of the task. Dependent tasks are coordinated in the overall schedule to set start and due dates for each task. People are assigned to manage each task, and team members are identified. Costs can be estimated (by task and in total) from the detailed schedule.

Planning Tools and Methods

Most project plans today are laid out on Gantt charts with supporting data in task lists, resource lists, cost estimates, and schedules. See Figure 5-1. There are a number of low-cost, PC-based, project planning software packages available to generate the charts and tables. These tools are just a convenience—project plans can be kept manually if the scope of the project is limited and/or there is an aversion to these PC-based tools. The thought and effort that go behind the

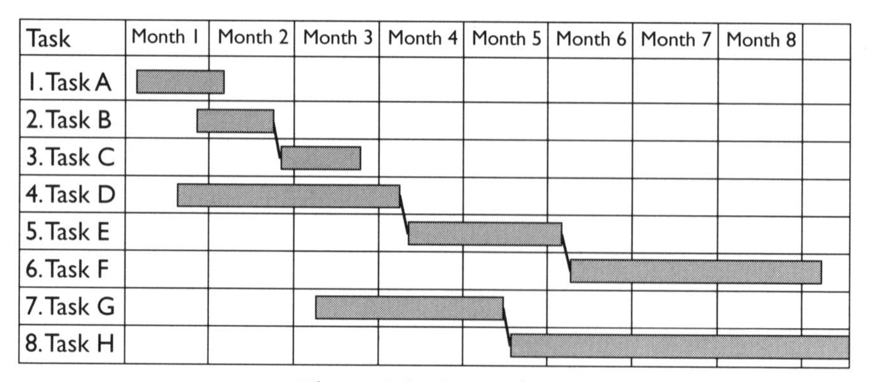

Figure 5-1. Gantt Chart

charts are the focus of this chapter. Select a method or a software tool (it doesn't really matter which one) and use it to organize the information and help keep the plan up to date.

An MRP II Implementation Plan

The high-level view of an MRP II implementation plan would identify the functional areas (or applications) to be implemented in rough sequence:

PHASE I: Financial
General Ledger
Fixed Assets
Payroll
Accounts Payable
Accounts Receivable

PHASE II: Customer Service
Inventory Management
Customer Order Entry
Sales Analysis
Forecasting

PHASE III: Operations
Product Data Management
Purchasing
Production Control
Bar-Code Data Collection
Engineering Change Control

PHASE IV: Planning
Material Requirements Planning
Master Production Scheduling
Capacity Requirements Planning

Please note that this is not a recommended implementation sequence or phase structure, but merely a sample for discussion purposes. In truth, I seldom recommend implementing the financial applications first, although it is common practice to do so. The financial applications are relatively easy to implement and can provide quick successes that encourage the rest of the team. They are not, however, the big payback applications. Phasing is seldom as simple as is illustrated. Usually, the application groups listed will overlap because of interdependencies and also because different people are involved in the different application areas and can be working in parallel with little interference. An equally valid plan might look like this:

	Primary Responsible
PHASE I: Customer Service	
Inventory (finished goods)	Materials
Order Entry	Customer Service
Accounts Receivable	Accounting
General Ledger	Accounting
PHASE II: Materials	
Inventory (raw materials)	Materials
Purchasing	Purchasing
Accounts Payable	Accounting
Bills-of-Material	Engineering
Material Requirements Planning	New function
PHASE III: Production	
Work Centers and Routings	Engineering
Production Control	Production
Bar-Code Data Collection	Production/IS
Engineering Change Control	Engineering/ Production
Capacity Requirements Planning	New (production)
PHASE IV: Planning	
Forecasting	Marketing
Master Production Scheduling	New function

Starting from this very high-level view, the next level of detail identifies more specific activities to be accomplished, as illustrated earlier when the inventory application was broken down into file loading, transaction procedure development, and so on.

This process continues until the identified tasks are small enough and manageable enough to be sized, scheduled, and assigned. By going back to the inventory example, the breakdown might look like this:

Inventory Management Implementation

Educate the project team

Educate/train task team members
Finalize the project and task plans
Install the software
Establish a standards committee
Set standards for codes, classes, units of measure, etc.
Load the item master data
Define and load the warehouse location map
Establish a cutover (transition) plan
Establish transaction procedures for the first group
Load balance data for the first group of items
Test/revise transaction procedures
Reload or correct balances (for the group) and cutover
Establish procedures for the second group
Repeat the steps for all groups (raw materials, subassemblies, etc.)
Begin cycle counting for the control group.
Establish procedures and begin cycle counting
Establish month-end procedures

For each identified task, there should be

- a task number or identification (for easy tracking)
- name and brief description
- a person responsible (task team leader)
- hours of effort required
- elapsed time required (based on hours and resources)
- planned and actual start dates
- planned and actual completion dates
- dependencies (item master must be loaded, terminals available before start date, users trained before start date, etc.)
- assumptions (number of people from a given department, on this task for X hours per day, system available during the time needed, etc.)

Estimating Hours

Perhaps one of the most difficult and risky parts of project scheduling is coming up with realistic estimates of the amount of time required to complete a given task. It is assumed that all or most of the activities will be new or changed from current procedures, and many will be one-time tasks with which the participants will have no prior experience. There may be a tendency to grossly over- or underestimate the time required and project schedules will suffer because of this.

One answer is to enlist the software vendor in helping you make these estimates. They have (presumably) installed this same software a number of times and should be in a position to provide advice based on this experience. They will not know the capabilities of your people, however, so they cannot make these estimates for you. Remember that new tasks and procedures must go through a "learning curve." You can simulate the process and measure the time it takes, but it is safe to assume that the speed for accomplishing a given task will likely increase with experience. If it takes two minutes to load an item master record on the first day, it might be reasonable to assume that by the second week it will only take one minute each.

Also watch out for the differences in people. If you try the task yourself and it takes two minutes per item, is it valid to assume that anyone else (the people who will actually be doing it) can do it in two minutes?

The estimating process relies on experience, knowing the capabilities of the people involved, gut feel, and common sense. If there are 2,000 records to load in a given file and your best estimate is three minutes each, including gathering the data, whatever validation is required (see what follows), and keying in the information, then the total time required is 6,000 minutes, or 100 hours.

Let's say that there will be two people involved in this task and that each can dedicate ten hours per week. At twenty person-hours per week, the elapsed time for the task should be five weeks (100 hours divided by 20). Schedule the task for a five-week duration. If other dependencies require that the task be completed in less time, additional resource must be applied. Get twenty hours per week per participant and the duration shrinks to two-and-one-half weeks. Add a third person at five hours per week and the duration is four weeks. This is the kind of project planning that is required to develop a properly resourced, coordinated plan. There is no rocket science here. The key to success is knowing the parameters—what must be done, how long it will take, how much resource is available, and what time frame is available.

You may find that there are some conflicts. Working down from the high-level objectives, you will develop a rough plan that will be the starting point for bottom–up detail scheduling. As the tasks are further defined and durations estimated, some will not fit or will require resources that are not available when needed. Some of the PC-based project management tools will help identify resource or schedule conflicts, but actual resolution of these situations falls upon management judgment. The project team will work these out as the schedule is developed. Be sure that you do the best job you can of developing a realistic schedule that considers all requirements and identifies and resolves conflicts before they can cause problems.

There will be surprises and there will be changes as the project proceeds. The trick is to keep them to a minimum by doing a good job of planning up front.

You may feel a tendency to "pad" the schedule, adding a little extra time or a little more resource than you really think you need. Resist. It is important to have an *aggressive* schedule. You are asking the participants to put forth extra effort and you won't be able to do this forever. The project schedule must be tight—allow enough time to do the job right but don't expect to be able to maintain enthusiasm and full participation for years. For the most part, people will respond to the request for extra effort as long as they can see sufficient benefit and the duration of the effort is reasonable. Padding each task in the schedule will drag it out to an unreasonable length and will also decay the benefits (payback), jeopardizing the business case.

Also, a padded schedule will become a self-fulfilling prophesy. If a task team is assigned a due date that is five weeks after the start date, the task will likely be completed in five weeks. If that schedule includes one week of "pad," it will almost certainly be consumed.

On the other hand, it is wise to allow a little room for error or unexpected delays. Pad the schedule if you must, but put the buffer *after* the task due date. This will protect dependent events from delays in earlier tasks but maintains the tight schedule for the tasks themselves. If all goes well, the dependent (later) task could be started early and, possibly, the entire project could finish ahead of schedule. What an accomplishment that would be!

Data Conversion

If you are moving from an existing computerized system to a new one, it is likely that some of your data files can be electronically converted. Before committing to an electronic conversion, consider these two things: If the file is not huge and can be manually loaded in a reasonable time, the experience of loading the data can be a great help in familiarizing the users with the procedure and the data itself. Make it part of the training. The second consideration is: How good are the existing data? One reason for moving to a new system is to improve management. Today's problems may be, at least in part, due to invalid, incomplete, or incorrect data in the system. If these data are simply rolled over into the new system, the problems roll over with it.

Conversion time is a golden opportunity for cleaning up bad data, but there is a counterconsideration: Data cleanup can be a major effort that will delay completion of the project. In fact, I have seen implementations where the cleanup task was so large that the system was not implemented. This is no place for perfectionists. Yes, you want the data to be as accurate as possible, but you cannot let the pursuit of excellence sink the project. Settle for as much cleanup as can be worked into the project schedule for a start, then follow up with a continuous improvement effort to make the data better through time.

To complete the electronic conversion discussion, do it for the large files where manual loading is impractical. Be wary of converting bad data—clean up major errors, but don't postpone the project in pursuit of perfection. For smaller files, consider manual loading as an opportunity for hands-on training and experience.

In some areas, the conversion might not be a one-time event. There may be an initial conversion prior to training, pilot testing, program modification and testing, and so on, with a reconversion just prior to "going live." In other cases, the implementation might be a phase-in process wherein certain product lines or departments or functions are converted first and others follow at a later time. This may require a phased data conversion. The whole subject of phase-in and parallel operation is addressed in a later chapter.

Task Team Leaders

Task teams come and go. They are assembled to do a specific job, and they disappear when the job is done. Each task team has a leader, whose name is on the project plan as the one responsible for completing the task on time. Ultimately, the project team is responsible for the completion of the entire project and all its tasks, but each task leader takes on responsibility for managing an individual task. The task team leader may be a member of the project team or task leader duties may be delegated. Tasks can be of any duration, but should not be too large. A task can always be broken down into several subtasks if it becomes too large to be easily manageable as a single task.

The task team leader should come from the business functional area affected by the subject of the task—someone from engineering to handle item definition, someone from materials to work on inventory balance. As stated earlier, the team leader must accept the assignment and agree to the resource plan and time schedule. If the leader doesn't have "buy-in," he or she will be hard-pressed to motivate the rest of the task team to do the job right and complete it on time.

If the Schedule Slips

Tasks must be completed on time. *This is critical.* The entire schedule is broken down into individual tasks, each of which fits in with the overall plan. If individual tasks are allowed to miss their schedules, the entire plan is at risk. Project team policy should be as follows: There is no excuse for an assigned task to be unfinished on its due date. Before you panic, let me explain further. The project team will meet weekly and the project is broken into manageable chunks (tasks), so visibility shouldn't be a problem. Once the task team leader has accepted the task and the schedule, he or she is responsible for completing it on time. The task leader will be immediately aware of any problems that might jeopardize timely completion. It is his or her responsibility to resolve the problem before the schedule slips. If the resolution is beyond his or her capabilities, the situation must be brought to the attention of the project leader and project team immediately—*before the task deadline.*

As a general rule, it is unacceptable to arrive at the due date and not be finished with the task. The task leader must notify the project team/team leader *before the due date* if there is a problem—early enough for the team to take corrective action to get the task back on schedule.

The team responds by doing what is necessary to complete the task on time. Usually, this means adding resources. If it is impossible to complete the task on time, the team may, at its discretion, change the due date, but that should be a *last resort.* Again, there are other tasks whose timing is tied to this task, and moving one schedule means moving all other dependent tasks, and their dependent tasks, and so on. Also, changing a due date sends a very clear signal to the rest of the project that due dates can be moved. As soon as the precedent is set, other tasks will become late and the entire schedule is unlikely to complete on time.

Procedures

When laying out the project plan and schedule, remember that the greatest impact of the system will be on changes to day-to-day operational procedures. You will be doing things differently with the new system—or the system will not have the desired effect (benefits). Remember the 50% accurate inventory system. When computerized, it became a 50% accurate computerized inventory system. What increases the accuracy is the change in procedures that reduces errors and adds validation and correction processes.

When sizing the implementation, recognize that new procedures must be developed as the new functions are incorporated into daily activities. Allow time for development of these new procedures and allow time for procedure documentation.

It is easier today to preach procedure documentation than it was just a few years age, thanks to ISO 9000. This International Standards Organization (ISO) quality standard is now being implemented worldwide and the heart of it is focused on documenting what is done and being sure that the documented procedures are followed.

Although a packaged software system will include system documentation (one of the big benefits of buying a package rather than writing the applications yourself), this documentation describes how the software works, not how you will use it in your company. Your procedure documentation may start with the vendor-supplied documentation, but must be extended to include local operational policies and how things work in your business: Where does information come from? Who provides it to whom? Who is responsible for entering the data? How are they verified? Who corrects errors and how? How is paper (audit trail) handled, stored, retrieved? When is it archived or destroyed?

In developing your plan and sizing the effort, be sure to include an allowance for this documentation task. The best people to do the documentation are those who will be doing the tasks. As they develop and/or learn new procedures, they should write them down. They should also maintain the procedures through time—as things change. You may want to use some standardized formats and issue guidelines and samples to help everyone produce consistent procedure documentation. Many companies store all procedures on a companywide network or server where they can be readily accessed and maintained.

The Project Plan

- List the detailed tasks needed to achieve goals
- Time-phase and coordinate tasks with each other
- Recognize dependencies
- Plan from the top down
- Tie to overall project goals
- Schedule from the bottom up
- Identify the effort required and resources available
- Schedule the time needed
- Assign each task a name (task leader)
- Assign a due date for each task (which must not be changed)

Managing the Project

Whatever plan and schedule are developed, the primary duty of the project team after developing the project plan is to manage the execution of the plan. As mentioned earlier, the project team should meet weekly to review progress, address problems, and look ahead to upcoming tasks. Team meetings should be limited to ninety minutes, if possible. If the meeting is well-organized, with a published agenda, and well-managed, this shouldn't be a problem. Ninety minutes (or less) should be enough time to handle things in most cases. It is important to keep in mind that the team members are busy people—key managers—and their time is precious. Keeping the meetings brief and efficient shows respect for the value of their time and encourages regular attendance.

Attendance at team meetings should be mandatory. You will need full attendance to ensure coordination and concurrence. If a team member is not available, a suitable substitute may fill in, but the substitute must have the authority to make decisions and commit the department he or she is representing.

A project team meeting should never be canceled because there is "nothing to discuss." I have tried to emphasize the need for an aggressive schedule in order to complete the project before enthusiasm runs out. An aggressive schedule requires continuous progress, so there will always be something to review and something to prepare for. If nothing happened in the past week, the project is in trouble and a project team meeting is certainly appropriate to bring the project back on track.

Management "Window"

In all likelihood, the overall project will extend for a year or more. Once the overall schedule is set and detail plans are made in the initial planning process, management of the plan will focus on near-term requirements. Many companies have been successful using a ninety-day planning window, that is, the project team will review and closely watch the plan for the next ninety days.

Let's say that the team is meeting on the first of January. The project management window extends from January 1 through the end of March (ninety days). The team will review the elements of the plan within that range, anticipate resource requirements, validate that everything is in place, and generally prepare for that portion of the project. At the weekly meetings during January, each week's accomplishments will be reviewed, any delays or problems addressed, and refinements made as appropriate. At the last meeting in January or the first meeting in February, the window will be extended through the end of April, bringing it back to approximately ninety days. Thus, the management team will always be focused on the next ninety days (more or less) as the project moves through time.

Excuses

Some tasks will not be completed on time. In some cases, there may be valid reasons, but the project team will have to decide which reasons are valid and which ones are not. I stated earlier that it is not acceptable to arrive at the due date and not be finished with the task. The task leader will know, before the task is due, if there is a problem. It is the task leader's duty to bring these problems to the attention of the project leader and/or project team as soon as they become evident. The team then can take action to bring the task back onto its schedule.

The important point here is to identify the problem and take corrective action before the task due date. Then, it may be possible to complete the task on time. If the problem is not addressed until the due date has arrived, it's too late.

It is common to want to "pad" the schedule to allow for the inevitable underestimate or unforeseen difficulty. By all means, do allow for the unexpected, but put the cushion *after* the task due date, not before. Hold the task leader responsible for completing the task on time. If there is an unavoidable delay, the cushion will allow dependent (later) tasks to start on time despite the delay in the previous task.

This may sound like fooling oneself, but it is really a valid management idea. Murphy's law says that work will expand to fill the available time. If the task duration is padded, be assured that the time will be used. If the extra time is not needed and it is placed (in the schedule) beyond the due date for the task, it is more likely that the task will complete on time and following task(s) could actually start and complete early. If the pad is used, there is minimal harm to the overall schedule.

You will likely hear the following: "I didn't have time to do ...," "I understood you to mean ...," "It took longer than I thought at first but now I'm catching up," "I had the time set aside to do this but then...."

The hardest one to deal with, and the most common, is probably that day-to-day business got in the way. When an employee is faced with a choice of completing a project task or meeting a production schedule, the production schedule will generally win. It's hard to argue with that. After all, production is the heart of the business and the project is "extra" and doesn't generate revenue or respond to the customer—directly. Unfortunately, there is no easy way to deal with this dilemma. The project is important and has sufficient benefit (return on investment) to justify spending the time and money to get it done. On the other hand, keeping the customer satisfied is the primary business of the company.

For the project to be properly planned and organized, this conflict will have been considered. The resource estimates should be made with this conflict in mind. The project is extra work, being added to the regular duties that the employees are currently doing, and will continue to do, during and after the project. The system *will* change what people do and should make them more efficient and more effective, providing more time for other things, but only after the system is implemented. During the implementation process, there is added work on top of what was undoubtedly already a full schedule. You will have to ask for extra effort, expect it, and get it to succeed with the implementation. Be sure the employees understand this and are properly motivated to give the extra effort when it is needed. They will only do this if they share your vision of how much benefit they will get—each one of them—from the new system.

Project Management

- Primary duty of the project team (after planning)
- Best done using a ninety-day window
- Each task must be managed to complete on time
- Expect conflicts with "regular" duties
- Make sure all participants know why they are doing the extra work (what's in it for them)

6. How Long Should It Take?

Just as it is ridiculous to make a career out of selecting a system, it is equally counterproductive to make a career out of implementing a system. Note that I'm talking here about doing these things at one company. Many of us have made a comfortable living by assisting many companies in selecting and implementing systems. But to tie up one company for any appreciable amount of time in either the selection or the implementation process is to delay the benefits that the system is designed to deliver in the first place. In fact, the longer it takes to implement, the less likely it is that the benefits will be achieved at all.

Don't laugh. Although it is usually not planned that way, there are many documented cases of perpetual implementations. A system can be implemented in a reasonable time frame and the sooner the better. But what is reasonable?

I've worked in the manufacturing business and planning system (MRP II) community for most of my consulting career and these systems tend to be among the most comprehensive and complicated packaged application sets on the market. Granted, not all companies try to implement all portions of a complete system, but overall I'd say that MRP II represents a typical, complex system that is a good discussion case for implementation duration.

The rule of thumb that consultants traditionally used as an estimated implementation time for a full MRP II system was eighteen to twenty-four months. People would always ask, and that answer was the one that automatically rolled off the lips—eighteen to twenty-four months. I, and many other experienced consultants, would only quote a duration reluctantly, because there are so many variables that there is really no such thing as typical. Why the reluctance? When quoted a range such as this, people only hear the number that's closest to what they want—in this case, usually the smaller one. Some implementations take longer than eighteen months, and I didn't want to set unrealistic expectations. On the other hand, if I said two years, the prospective implementer might reconsider the entire project because such a long implementation is really hard to commit to or maybe even to contemplate.

So, eighteen to twenty-four months was the default answer and it was a reasonable range and an achievable goal if the company is serious, committed, and organized.

I use the past tense here because that rule of thumb is not as reliable as it used to be. All else being equal (whatever that means), eighteen to twenty-four months for full MRP II implementation is a reasonable expectation. This assumes that it is "typical" company with little or no system support in place, well-organized and motivated, and the goal is to implement the usual suite of financial, operations, planning, and customer service applications and be using them to manage the business.

Identifying the end point is one of the basic problems in setting an implementation time frame. Installing MRP II or any other comprehensive information system is a journey, not a destination. There will be many points along the way where subsystems or functional areas of the business will be fully "up and running" and the project could be possibly considered complete at any of these junctures. On the other hand, no system is really complete because there is always more that can be done.

A "typical" complete MRP II system will include the basic operational applications (inventory, product data, production control, purchasing), financial applications (general ledger, accounts payable and receivable, possibly payroll and fixed assets), customer service (order entry, invoicing, distribution, sales analysis, forecasting), and planning (material requirements planning, capacity requirements planning, and master scheduling).

If the goals are more modest, for example, not including some of the planning and omitting other than basic production control capabilities, then the time frame could be shorter. If the company is larger or the products or processes are complex, the implementation might take longer.

I reiterate that MRP II is a journey, not a destination. By the time the initial objectives have been achieved, there should be a whole new set of objectives waiting in the wings. As you absorb each portion of the system (each application or functional area of the company, for example), the accomplishments to that point become a foundation upon which other applications or other areas of the business can be addressed. You may start with basic product data and inventory management as the first phase, then add purchasing and production control, then go to automated data collection and basic planning, then advanced planning, then distribution planning and management, electronic data interchange, maintenance management, integrated quality management, and so on.

Each "phase" of the process should have its own objectives and time frame. A reasonably comprehensive basic implementation with the usual dozen or so applications will typically end up being a one- to two-year project.

There is also a lot of psychology inherent in a system implementation project. When I first started out as a consultant, I used to think mostly in terms of the technical challenges and the details of software functionality, fit of the package to the situation, the mechanics of laying out a plan, and the like (I have an engineering background, after all), but the longer I worked in this business, the more obvious it became that the people issues are the important ones and therein lies the psychology.

Let's say you are the project manager for a system implementation and it is your job to assemble the team, draw up the plan, and shepherd the project through to completion. Let's also say that the project is estimated to take four years. Imagine standing in front of the assembled implementation team and explaining to them that they will all be expected to attend team meetings, go to training classes, organize and supervise subteams and accomplish specific project tasks, load data, change operating procedures, and generally give their all, *in addition to doing their regular jobs*, and that this extra work would drag on for four full years. Now imagine the deluge of résumés that would be literally pouring out of your company in the next few weeks and months.

A system implementation is extra work—there is no doubt about it. Virtually all employees will be impacted by it (if it is truly a comprehensive system) and most will be asked to contribute time, effort, and emotional energy in the process. If the benefits are properly communicated, there is no reason why loyal, forward-thinking employees cannot be convinced that it is in their best interests to participate (the "what's in it for me?" factor) and assist. But it is unreasonable to expect this extra effort over a long (or endless) period of time.

Furthermore, it helps to divide the project into smaller pieces, each with defined tasks and benefits, so that commitment and participation can be developed in more manageable chunks. There must be intermediate measurements: a series of small victories to make it easier for the participants to grasp the requirements and keep their enthusiasm over an extended period of time.

Once we agree on what "implementation" means, there are still the many variables from company to company that will affect the duration of the project. Earlier, I mentioned a "typical company with little or no system support in place, well-organized and motivated." Typical is a dangerous word. Companies will vary in size, complexity, the kinds of products and the markets they serve, processes, management style, and a thousand other ways. Some of these variations will affect the speed and duration of the implementation process.

If the company already has an MRP system in place, that may speed up the implementation of the new system. The users will already know what MRP is about and what it takes. Most of the required data will be available for electronic conversion to the new system. Many of the required processes and procedures will be in place. Therefore, implementation should take less time. UNLESS—the incumbent system is a failure and there are bad habits and poor attitudes to overcome. UNLESS—the users love the old systems and resist the change. UNLESS—the new system is so different from the old one that the users find it hard to make the changeover. UNLESS—there is a long period of parallel operation (more on this later.) UNLESS....

As you can see, there are many variables.

Lately, software vendors have been trying very hard to add features to their products to make them easier to learn and easier to use. Foremost among these usability enhancements is the use of what's called a Graphical User Interface, or GUI (pronounced Gooey). Generally, this uses the Microsoft Windows® formats (look and feel). The theory says that many of us now know how to use Windows and Windows applications like spreadsheets and word processors. If business applications look the same and work the same way, it should be easy to use them. Some vendors talk about "walk up and use" applications. Don't believe it.

First of all, there are still many people who aren't familiar with Microsoft Windows. Second, I don't believe that Windows is that "intuitive." Some people claim that Windows (and therefore Windowslike applications) is so obvious that no training is required. Someone completely new to the application should be able to figure it out with no help. I don't believe it. My own experience shows that knowing the interface (screen layouts) only goes so far in understanding how to use the application.

In addition, the operation of the software—which keys to press or which icons to click—is not where the challenges and the benefits of the system are. It is the application of the systems capabilities to the business problems at hand that yields business results. Knowing how to push the buttons is a long way from knowing how to apply the functionality to your business.

In any case, software vendors now claim that their systems are easier to use and easier to learn and some of them promise shorter implementation time—six to nine months, for example. This is certainly attractive, if true, because the sooner the implementation is complete, the sooner all those benefits become available. Be careful, however, in blindly accepting these short implementation time promises. I'm not convinced that *any* packaged software can be so intuitive or easy to use that the normal implementation cycle can be cut in half. Easy to use is only a small part of the challenge.

At the opposite end of the scale, there are vendors who like to sell packaged software solutions bundled with lots of consulting services. These situations can be far more than simple system implementations; they often include business process reengineering or similar extensive consulting and business restructuring efforts. A cynical view is that the software is just a vehicle for selling consulting services. This kind of project often lasts much longer than the typical packaged software implementation and the vendor is generally more interested in selling the services than the package. It is not unusual in these cases for the total project costs to be five to ten times the cost of the software. Some of these projects last as long as five years or more.

How Long?

The implementation project will take as long as is needed to install the hardware and software, train the users, establish new procedures, and make the system a part of the day-to-day business of the company. If properly planned and managed, this time frame should be reasonable—meaning long enough to get it done, but not so long that it is a burden on the users and the company.

In most cases, a reasonable project duration is twelve to twenty-four months. If a viable plan can be carried out in less time than that, all the better. If your working estimate is greater than twenty-four months, you've got a problem. Either tighten up on the schedule or revisit (cut back on) the objectives.

Initial project planning is a top–down process. Identify the goals and when you want to be there, then take a stab at how to get from where you are to where you want to be.

Project planning is more bottom–up. Once the individual tasks have been identified, each is sized and scheduled. The duration and sequencing of project tasks will determine the actual schedule and duration of the entire project. Chances are good that the bottom–up schedule will not match the top–down plan. Refining the plan and the schedule to create a viable project is one of the key jobs of the project team.

Implementation Project Duration

- The default MRP II implementation time frame is eighteen to twenty-four months.
- Duration for a particular project varies with company background, extent of the project, attitudes, experience, and so on.
- A system implementation that is too large (too long) can be broken into a series of smaller projects, each with a duration and end point.
- The project team develops the detailed project schedule and coordinates it to the overall project objectives.
- A reasonable project duration is twelve to twenty-four months.

7. Organizational Impact

I've heard and read a lot lately about silos or chimneys—some people call them pipes or stovepipes—in reference to the structure of a manufacturing organization and also its information systems. It seems that the popular view is that traditional MRP II systems are organized around so-called vertical functions and this organizational structure inhibits the free flow of information *around* the organization and between functional areas. Horizontally, so to speak.

I can see their point. If you look at a modular MRP II package, you will see an engineering module (product data management or some such), an inventory module, a production control module, and so on. The impression might be that each of these modules addresses the designated functional area without concern or consideration for other areas. See Figure 7-1. That impression would be dead wrong. The very essence of MRP II is its integration. Production control cannot operate without the support of engineering and inventory, planning relies on information from virtually every other area of the system, customer service interfaces directly with material control and planning and production control, and on and on.

There is a common problem, however, related to the modularization of packaged software and the way it's implemented. Because of the functional packaging, we tend to organize the implementation effort around these defined areas. As a result, we sometimes end up with a reinforcement of the interdepartmental barriers that are traditionally present in a manufacturing organization (or any other kind of organization for that matter). Another unfortunate by-product of the functional modularity of packaged MRP II is that built-in

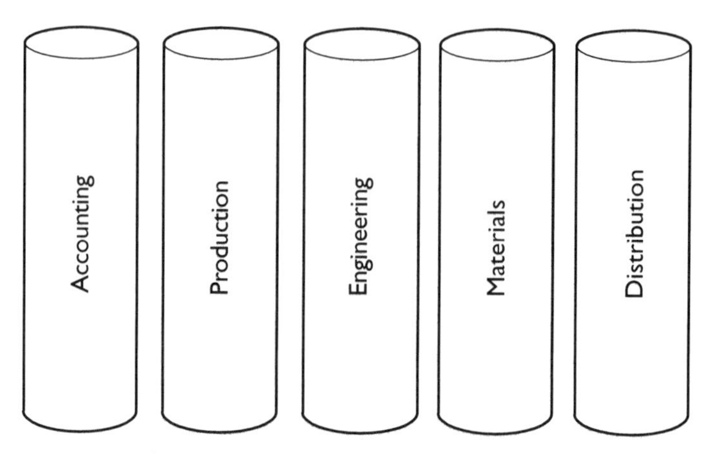

Figure 7-1. Functional Silos or Chimneys

inquiries and reports tend to stay pretty much within the bounds of the functional area; the inventory application contains inquiries that focus on inventory data, production control inquiries and reports deal only with production control data. Users can be given access to inquiries in other modules, but this is often discouraged because of proprietary interests and general fear of what these "outsiders" might find when they are looking around in *our* data. The new crop of Graphical User Interfaces (GUIs) helps alleviate this situation somewhat by making inquiries and maintenance access to data less application-dependent.

When I present MRP II topics to an organization that is about to install such a system, I always focus on the integration, which is the strength of MRP II, and the need for communications between rival groups (departments) within the company. MRP II doesn't work unless these barriers are torn down and employees feel free to openly exchange information. I like to say, "you can't hide any more" and "there can be no little empires in an MRP II organization."

Unfortunately, the culture of an organization, its personality, doesn't change overnight. Without strong leadership and a firm commitment from the highest levels of the organization, the project is severely limited in its potential for success.

In one of the finest examples of MRP II in action that I have had the pleasure to observe, the company went to great lengths to open up the database to all users. In addition to the standard inquiries provided by the software vendor, they wrote many more programs to combine information from other "modules" into convenient lookup screens and combination reports. I actually saw a customer service person first check the standard inventory and customer order records while researching a customer question, then switch over and look at the current shop schedule, the location of the job, and the projected completion date. Not many production managers are willing to let customer service see their schedules and priorities and pass this information directly to customers.

Obviously, this company has invested heavily in the tools of organizational openness and also in user education. The customer service agent not only had to know how look up this information, but also how to be able to interpret it and understand what the customer should be told under various circumstances. I'm not advocating that anyone lie to a customer, but there are some things that can be revealed openly and others that should be considered internal information only. If people are allowed access to information, they must also know its value and how to protect it.

In any case, the stovepipe or chimney view has some basis in fact, but only because of the nature of people in an organization and a limited view provided by the packaging of software systems.

It is also true that earlier systems may have been severely limited in the ability to grant access to information across modular boundaries. With today's MRP II, built around a relational database management system (RDBMS), there's really no system-related excuse for packaging data into functional chimneys. Open access to data is easy to implement with an RDBMS-based system and it is principally the outdated ideas about function and responsibility that prevent wider access.

Fourth-generation software tools allow easy access to data in an RDBMS and many companies grant users access to query-type utilities for ad-hoc access to information. Of course, any tool can be abused. Users must be taught how, when, and under what conditions the tools should be used and strong incentives should be in place to protect the database and preserve system responsiveness. In a well-managed environment, however, there is much to be gained by expanding access to information.

It is generally agreed that one of the prime aspects of competitiveness today is time. A company that can respond faster to customer requests, a company that is faster to market with new products, a company that can detect changes in demand and respond quickly has a great advantage over its competitors. All of these things are information-driven. Allowing more brains to have more access to more information increases the odds of detecting and properly responding to changes.

The only function of MRP II or any other computer/software system is to manage information and make it available to the users in useful formats. You could say that a system turns data into information—something that has value. The wider the distribution of that information, the more valuable it is likely to have. Don't let your thinking be limited by marketing-driven software packaging compartmentalization. Think beyond traditional departmental boundaries in your own organization.

Knowledge (information) is power. Share it.[1]

To expand a little further on this idea of compartmentalization and interdepartmental rivalry, I believe some of this competitiveness stems from the way we allocate resources. Capital budgets are limited. When budget time rolls around each year, each group (department) will have needs and wants for the coming year and each knows that not everyone will be satisfied. Thus, departments compete for these limited resources. If engineering gets its new design system, chances are good that production will not get the new milling machine they asked for.

This rivalry or competition pervades the organization, not just at budget time, but year-round. When an order is not ready on time, distribution (customer service) might blame production. Production will claim material shortages (purchasing) disrupted the schedule. Purchasing might try to pass the buck to inventory: "The balance records were wrong. They said there were plenty and there really weren't any" And so forth.

Because departments are separately managed, and each department is measured on supposedly isolated factors that are seen to be within its control, the interactions between departments become battlegrounds.

---

[1] The preceding is adapted from my July 1993 column in *3X/400 Systems Management*.

The organization that you now have was established, or more likely evolved over a long period of time, to support the way you currently do business. Assuming that the implementation of a new information system will be accompanied by changes in the way you do business, as it must, then there is reason to suspect that some organizational changes will also be required.

At the very least, the system will automate some functions that are now handled manually, especially clerical tasks like typing and filing. But it will also create the need for new positions, and may require that some reporting structures change to accommodate changing responsibilities and changed lines of communications.

Tasks/Positions Eliminated

Since the system will undoubtedly take over or simplify the tasks associated with storing and retrieving information, some of the labor required to do those things manually will no longer be required. Don't assume, however, that this means that the overall net labor required will necessarily decrease accordingly. There will be new tasks associated with the care and upkeep of the new system's information management capabilities that will replace a good deal of the labor that is eliminated.

It is a grave mistake to implement a system with the expectation of massive head-count reductions. Not only is this expectation unlikely to become reality, but company employees will be understandably reluctant to embrace a new system that could well cost them their jobs (see Chapters 8 and 10). Nevertheless, you will likely need fewer filing clerks and less general clerical help when you have a comprehensive information management system in place and functioning.

Perhaps surprisingly, data entry positions in the data processing (information services) department should disappear. Some data capture may be simplified or eliminated through advanced technology, but the largest impact will be the distribution of data entry responsibilities out to the user departments. It is essential that the users take ownership of their portions of the system and a significant part of that ownership involves taking responsibility for the data. If your current systems department is providing data entry services to the user community, a part of the changes brought about by the new system implementation should be to end that service. The data entry clerks may be transferred out to the user departments (data still must go in and the users probably don't have the time to add this task to their daily duties without more help). There will likely also be some realignment of duties so that the data entry people can become more a part of the departments they are assigned to. Be prepared for the head-count reduction in IS and the increase in user department head count. Net change: probably zero.

There's also the possibility that intermediate levels of management may disappear. The great "flattening" of the corporate organization that has been occurring for the last few years is supported by the increased availability of information and enhanced communications engendered within today's integrated information systems. Although these are true personnel reductions, they cannot be credited (or blamed) solely on the system and it is important to make that distinction. If the system is seen to eliminate jobs, people will be reluctant to support it.

Other jobs will be eliminated due to better management. A prime example is expediters. Expediting is taking extraordinary action to shorten the normal or required lead time to acquire purchased materials or to accelerate the completion of production activities. Expediting is required when things go wrong—bad forecasts, promising delivery to a customer in less than the time required to complete the job, last-minute schedule changes (brought about for any of a number of reasons), unexpected rejects or scrap and more. Expediting is expensive and disruptive to the flow of activities in the company.

For many companies, expediting is a way of life. One company I worked with a few years ago had seventeen people on the payroll with the title expediter. And these were not low-paid clerks. If there were seventeen full-time expediters, I can guarantee you that there were many, many more people in this organization who spent a significant portion of their time expediting—often referred to as "chasing parts." This includes buyers, production managers, supervisors, and others.

And all of this time and effort is waste because it is done only to correct situations that could and should have been avoided. If the new information system is used to help get organized, to plan better, to communicate more effectively within the organization, and to anticipate problems so that they can be avoided, then the need for expediting can be greatly reduced.

The talent, experience, and intelligence of the former expediters can be applied to more rewarding and more effective tasks as part of the new orientation toward planning and control. Some head-count savings may occur with the elimination of wasteful practices like expediting, but before simply cutting the manpower budget, think about how the talent and experience of these "doers" can be exploited for more positive results.

When tasks disappear, don't expect to see people standing around or napping at their desks. Murphy's law states that work will expand to fill the available time, and Murphy knew what he was talking about. Few of us will voluntarily admit that we don't have enough to do. In fact, just about everyone has too much to do and any relief simply makes room for duties that weren't getting their proper attention before.

And don't expect anyone to come to you and say, "You don't need me any more." As tasks are eliminated, people will find other things to do that will become more important in their minds as they focus more attention on them. Some of these new duties will not be important to the achievement of company goals, even though they might sound like noble pursuits and will be staunchly defended by the person(s) doing them.

As tasks or positions are eliminated, have a plan for where and how the newly freed resources (people and man-hours) will be applied. If you let the workers find their own things to do, don't expect the company to benefit from the elimination of some work or duties. This sounds extremely cynical, and I certainly don't mean to imply that all workers will automatically choose nonvalue-adding busy work to fill the hours. As project leaders and managers, it is important to view the availability of resources in context. If a given task will be eliminated or reduced, figure out the best place to use the resource (person(s), space, equipment, etc.) to advance the goals of the company and the project. Be prepared to provide retraining or other assistance to the employee whose duties are being changed or eliminated. Keep in mind that the employees' experience is a valuable resource that must not be squandered or discarded. New job skills can be learned easily. Knowledge of your company, its products and market, personalities, values, and processes is much harder to come by.

New Tasks/Positions

Implementation of an MRP II system will add positions at a number of levels. Although filing and clerical tasks will be eliminated, there will be new requirements for data entry and data management (validation, auditing). Life will be easier in the stock room because the system will help locate materials and keep better records of what is available. There will be more and better information to support more efficient picking, put away, effective use of space, and so on. On the other hand, there will be a greater need for data entry and retrieval that replaces some of the manual record keeping. New disciplines and a greater focus on accuracy and data integrity will add the need for data validation tasks such as cycle counting and error investigation and correction activities.

Instead of using expediters to make up for lack of planning or control, there will be a greater need for planners and schedulers. It may look like these new tasks will require an increased head count, but new efficiencies and the elimination of unnecessary activities like expediting should free up sufficient resources to fill these new positions. During implementation, there will be a period of time when the new duties (master scheduler and material planners, for example) have started, but there has not been enough time to reduce or eliminate the expediting and manual scheduling. Factor this into your plans. There's no easy answer—there will be extra work for a time during the implementation process. Just don't assume that the added work load is permanent. Whereas the new duties will remain, some of the old tasks and duties will eventually go away, if you are successful in carrying the project through to completion.

This is an argument for an aggressive implementation schedule. New duties are introduced before efficiencies eliminate the old ones. The work load will definitely increase before it decreases. The longer the implementation (transition) period, the longer people will be required to work harder and longer. Eventually, they will tire and it will become harder and harder to keep them fully involved and committed. The longer the implementation project continues, the less likely it will be to bring it to a successful completion.

It helps to have some baseline measurements and savings estimates before the project starts. The visibility thus provided supports the need to identify the full cost of the project (including extra effort and the postponement or elimination of other duties), and planning for changes in work load both during implementation and after, as discussed before. Work with the department heads and supervisors to get agreement on what the staffing and the duties will be like after project completion and work together to drive toward those changes.

The system implementation had progressed very well so far—the basic applications were installed and the users were starting to feel comfortable with the new procedures. At the regularly scheduled executive steering committee meeting, Henry, the project leader, made his report. After relating the good news, the progress and successes, he stepped into what he knew would be a controversial area:

"Now that we have inventory under control, the next application on the schedule is MRP," he announced, pointing to the Gantt chart. "The team has identified who they think should be the new planners and a basic MRP class is scheduled for the week of the tenth of next month. Here's the proposed organization chart."

There are a few moments of silence as the executives study the chart. Then the production manager realizes that Sally Jones, his sharpest scheduler, will be moved to the new planning department. "You can't have Sally," he says, "she's the only one who can handle marketing's last-minute specials."

Don't expect department managers to openly and gladly accept all of these changes. Managers tend to have a proprietary feeling about their people and the duties they perform. Many will object to losing either one. On the other hand, expect a reluctance to accept new duties, especially if there is no increase in head count. The executive steering committee will want to make the big decisions about duties and staffing and see that its wishes are carried out.

Other Organizational Changes

Successful implementation will change the communications flow within and throughout the company. As an example, let's look at how a buyer might spend her day without benefit of an integrated system.

- Work through the "shortage list" to expedite needed materials.
- Take dozens of phone calls checking on the status of open purchase orders.

- Call vendors asking about status of orders.
- Take more phone calls requesting new purchases.
- Spend hours at the files (or have an assistant help) looking for receipt records and QC reports, looking for purchase history, researching sources, tracking open orders and late receipts.
- Handle numerous paper requisitions.
- Type (or have assistant type) new purchase orders and revisions.
- File multiple copies of purchase orders and distribute copies to receiving, QC, the requester, and accounting.
- Take more phone calls from people who want to know sources, prices, delivery lead times, etc.
- Maybe, if there's time, research potential new suppliers, analyze vendor performance, negotiate, and manage.

With an integrated system, most of the phone calls can be eliminated because the information is in the system and available to anyone who needs it (and has authorization to access it). Instead of calling for purchase order status, the production or materials people can call it up on their own screens. The same for prices, lead times, and so on. With tighter control and increased visibility, there should be less need to call the vendors for status information.

Because the system plans better and coordinates activities better than are possible manually, the shortage list should be shorter to begin with, and the system will recommend corrective action so there's less research required. Requests for new purchases will come through the system, either through MRP recommendations or via electronic (paperless) requisitions.

It goes without saying that filing and retrieval are a lot more efficient on a system. Most of the multiple copies of purchase orders can be eliminated because information is available on-line, throughout the company. And, finally, the system will print purchase orders and requisitions, so typing is unnecessary.

The majority of the time savings center around changes in the methods and the need for communications with internal "customers" and external suppliers. External communication is enhanced even more through electronic communications such as Electronic Data Interchange (EDI), a standardized protocol for exchanging business documents (purchase orders, invoices, advance shipment notices) computer to computer.

How can the buyer use all of this newly recovered time? Doing the things that she is really being paid to do: research potential new suppliers, analyze vendor performance, negotiate, and manage.

Standing Committees

Because things must be better coordinated between departments with an integrated approach, it is highly recommended that standing committees be established to promote this coordination. A standing committee is one that is perpetual—meeting on a regular basis to accomplish a continuing duty or task. There are also some specific system requirements that these committees can address that are important to successful operation.

The project team is the first of your working committees, but its job is of limited duration (the life of the implementation project). But, think in terms of continuous improvement. I have said before that system implementation is more of a journey than a destination. The system implementation itself has a number measurement points (milestones) on the way to the big goal, which is probably a twelve- to eighteen-month proposition. By the time the big end point is approaching, there will undoubtedly be extensions and/or other projects that have been identified that will expand on the accomplishments of the original project. The working relationship established among the team members during the implementation project can be preserved and applied to add-on projects.

Because all applications are established around a single, comprehensive database, there will be certain pieces of information that must be consistently entered and coordinated across the company. Examples include unit-of-measure codes or abbreviations, item classification groupings, abbreviations used in describing items (to facilitate alphabetic searches), naming conventions of various sorts, and the like. A standards committee should be established early in the project to start developing these guidelines and lists, so that the data are loaded consistently from the beginning. This group should remain in existence beyond the end of the project, however, because things change—there will be new products developed, new uses of the system may dictate changes or additions to the naming conventions, and new applications will introduce additional codes and fields to be coordinated.

Usually, a costing standards committee is also warranted. Most comprehensive systems will tend to distribute portions of the costing duties by virtue of the fact that costing is an integral part of operational activities, not a separate one. Product costing (standard costing) will rely on the bills-of-materials and item definitions that are the responsibility of the engineering group. Guidelines must be provided to engineering so that its work will support the needs of the costing system, and not work against it. Actual (job) costing functions will assign values to operational activities. Production must be aware of any requirements placed upon the way it sets up its tracking system and report activities in order to collect the correct cost information. And the committee will work continuously to develop improvements in how costs are collected and evaluated both to support its own activities and also to provide useful cost-related information back to the operational departments.

Consider a data integrity task force or committee as a type of quality control function for your systems's database. This group would search, analyze, and test the data that support the system's functions in an effort to detect problems or errors before they cause trouble. It will look for consistency between related or relatable fields (if the item is defined as a sellable product in one field, then the planning control should be X or Y and the order policy should never be Z). The task force can also identify potential improvements in data entry or maintenance procedures that will enhance accuracy in the future. Fourth-generation database tools that allow ad-hoc access to raw data are especially useful for this kind of work.

Most companies also require a configuration control or change control committee that meets regularly to coordinate on product definitions—what the official content is for each version or revision of each product and what (if any) variations are allowed.

Another standing committee, one that probably already exists in some form at your company, is the group that decides on the handling and disposition of nonconforming materials, parts, and products. Often called a Material Review Board (MRB), this group decides what is returned to vendors, what is reworked and how, if and how defective parts or materials can be used for other purposes, and how to dispose of unusable items.

You may think of other standing committees that would help coordinate the activities of your company and assist your efforts toward further refinement and continuous improvement.

Organizational Changes

- Anticipate and plan for the changes in the organization.
- Prepare for structure and organization changes, although overall head count may not change much.
- Expect new duties to begin before improvements can eliminate tasks/duties.
- Work with department heads and supervisors to anticipate and manage these changes.
- Provide sufficient retraining for reassigned employees.
- Establish standing committees to set up and maintain standards, data integrity, change control, improvement, etc.

8. Changes Large and Small

Last year, I read a column in the *Boston Globe* that really caught my attention. The columnist wrote about the tendency in his household to resurrect old ideas long after they had been overtaken by events. For example, there was a fire in the family garage that damaged the roof. In a discussion about getting it repaired, the columnist's wife remarked: "Maybe this would be a good time to add the room over the garage for Junior like we've always talked about. They can do it while they are fixing the damage."

Well, it turns out that "Junior" was grown and gone by this time, but the idea of adding a room over the garage had embedded itself in the subconscious mind and poked out into the open when the garage roof was the topic of discussion. This kind of thing had happened in this family so many times before that it had developed a standard response, a kind of a catch phrase to recite as an automatic response: "That's last year's war."

Things change, but thoughts and ideas have a way of lagging changes in the facts. Once a thought lodges itself in the brain, it can sometimes be impossible to shake it loose, no matter how inappropriate or obsolete it has become. This is perhaps a manifestation of the widespread reluctance to change. We tend to cling to the familiar and resist anything that threatens the status quo.

In a dynamic environment, such as is found in most businesses, inertia can be deadly. Recent history is littered with the carcasses of companies that did not respond to changing conditions or responded too late. Those that have not been put completely out of business are mere shadows of their former glory. Look at Wang computers, once the undisputed champion of the word processing world and a major player in midrange office systems. Wang management failed to recognize or respond to the movement of word processing from dedicated systems to PCs and it also underestimated the market appeal and capabilities of the IBM midrange product line. Wang's response was too little too late. It continued to produce uncompetitive midrange systems and completely missed the switch to PCs. In Wang's heyday, who could have predicted the drastic downsizing and staggering losses? On the other hand, after the market and technology changes became obvious to even the casual observer, who would have predicated that Wang would stubbornly cling to its sinking ship until it was far too late to recover?

Wang is certainly not alone. The history of business is littered with the rusting hulks of companies that failed to recognize market changes. Some were able to recover, albeit with great difficulty and almost invariably as also-rans in markets that they originally created and/or once dominated. Established, comfortable organizations are like secure, settled individuals. We all tend to get "set in our ways." Hardening of the attitudes can be as lethal to an organization as hardening of the arteries can be to a human body.

To make the point a little more specifically, let's look at a typical production management methodology. Despite the widespread acceptance of just-in-time (JIT) principles, many direct production workers are measured and rewarded for quantity of product produced. This encourages overproduction and ignores priorities and direct responsiveness to market demands. Even in companies that espouse JIT and related management views, many still watch and worry far too much about traditional efficiency and utilization measurements. They put elaborate planning programs in place and then destroy their effectiveness by measuring the wrong things and thereby giving the wrong signals to workers on the shop floor. It seems that these traditional measurements are just too solidly ingrained to abandon.

What's the cure? It starts with awareness. The healing process can begin only when we realize that there is a problem. Just as the newspaper columnist and his family earlier developed a catch phrase to point out obsolete thinking, we must all find a way to call attention to potential lapses.

Encourage an attitude in your organization of constructive questioning, especially when decisions are made by default. What I mean by that (decisions by default) is when things are done the way they have always been done, without thought about whether that is truly the best way. Just because it made sense last year (or last month) does not necessarily mean it is still the best way or even that conditions are the same as they were when the procedure was developed.

Become an organization in touch with today. Encourage your co-workers to question the way things are done by asking open questions such as: "Why do we ... ?" or "Does this seem right to you?" Try to avoid confrontational questions such as: "Why did you ... ?" or "Shouldn't you be doing ... ?" See the difference? Approach each situation as a partner, not as a critic or an accuser. Question your own processes and procedures as an example to others.

Remember, too, that many people avoid risk as much as possible. Change and questioning the status quo represent risks. Teamwork helps dissipate fears by spreading risk and offering comfort and encouragement. Whenever possible, use teams to assess processes and recommend improvements.

Responding to changing conditions (changing markets, changing technology, changing management style or direction, changing character and availability of information such as from an information system change) can be accommodated continuously, as a series of small incremental changes or all at once, in a cataclysmic upheaval. A continuous process of small changes may achieve the same results as a period of no change followed by traumatic upheaval. Think of the San Andreas fault. It is dangerous because stresses are not being relieved by small, incremental earthquakes. When the "big one" comes, there will be hell to pay because all of the change happens at once. Wang and many other companies have experienced their own versions of the "big one." Whether you choose continuous change or incremental change, change is a necessity. Markets, products, and technology do not stand still. To keep up with the world, an organization must be willing to adapt.

Reengineering

In the business world, the functional equivalent of the major earthquake is now called reengineering. There has been much attention paid to reengineering in the press in recent years, and it all started with a 1990 *Harvard Business Review* article written by Dr. Michael Hammer, a Boston-based consultant. In "Reengineering Work: Don't Automate, Obliterate," Dr. Hammer declaimed simply automating existing processes. He feels that companies should not necessarily feel bound to the equipment and technologies that happen to be currently installed at their plants.

The idea of reengineering is akin to the old philosophy of zero-based budgeting in that the designer (or budgeter) should not advance the present situation with incremental changes, but rather should justify each future planned action on its own, disregarding limitations or even suggestions based on past decisions or investments.

Conceptually, a reengineering project should result in an optimum (ideal, philosophically) solution, whereas incremental change will nearly always be a compromise between what is now in place and the ideal. In the pure reengineering ideal, it's a clean sheet of paper. Imaging blowing up the current facility and starting completely new—what is the best way to do this job? A reengineering project, therefore, will almost always result in a more expensive and disruptive solution. Few companies can afford to essentially toss out all past investments and start over.

Reengineering, like most of the "buzzwords du jour," gets a lot of attention, spurs the growth of the consulting industry, and is actually practiced by few. Also like most other "new" ideas, there are remarkable successes and notable failures to cite in the trade press and through the ever-present grapevine. Like most of the others, too, we can learn much from these new ideas or approaches and apply them, if not in total and as presented, in our efforts to keep up with a changing world.

I believe the best lesson we can pick up from reengineering, whether we can use it (reengineering) directly or not, is a reminder to expand our thinking when we are looking for ways to improve processes, gain control, and meet the competition. Many times, we address individual problems or situations with solutions that are focused strictly on the specific problem. Often this leads to a symptomatic solution rather than a systemic one—treating the symptom rather than the disease.

In many cases, solving a problem in one area of the business can create problems for other areas. In an effort to reduce inventory, for example, operating procedures in the material management area might be changed to reduce the lot sizes for incoming materials. Smaller lot sizes means more orders. This puts an additional burden on purchasing, to release and track more purchase orders, and receiving will have more receipts to process.

Of course, there is danger in going too far with a "clean sweep" approach. If our thinking is entirely unrestricted, we may end up with a solution that ignores what good things there may be in our current processes. How can we take advantage of burden-free thinking yet not ignore the assets and talents we already own? We take advantage by a three-part approach incorporating the best of reengineering and the traditional incremental improvement approach.

Step one would be to design the unfettered ideal solution. Using the reengineering idea, design the systems and processes that are the best fit using current and emerging technology.

The second step is to identify what is good and bad about the systems and processes in use today. Good and bad in this case refer to appropriateness and compatibility with the ideal solution developed in step one.

In step three, we engineer the "practical ideal" solution, starting with the best of the current and identifying the most practical path to replacing inappropriate systems and processes that will move us toward the ideal.

Continuous Change

Continuous improvement is more akin to the way most organizations normally institute change, although for many, the changes are not always improvements and the improvements are not always continuous. Nevertheless, in the absence of something like a reengineering effort, change takes place gradually and incrementally. Most often, projects are initiated to address a specific problem or implement an improvement aimed at a specific area of the business.

Continuous improvement efforts may take the form of a daily flow of ideas and changes that are implemented individually as they come up and are deemed worthy. Many Japanese companies use such an approach, which they call *Kaisen. Kaisen* can be defined as a process of small, incremental changes that are driven by suggestions that come from the workers, operating as self-directed teams.

On my study-tour of Japanese industry a few years ago, I was very impressed with the fact that all the companies we visited actively encouraged suggestions and were very proud of the improvements that had come from these worker-teams. This comes about through a basic philosophical approach to employees and how they interact with management and the company as a whole, and also the way employees are organized and compensated. In Japan, the group or team is all-important. In the West, we admire independence and individual achievement. In a team-oriented framework, each member of the team feels responsible for the accomplishments of the entire team—individuals are not encouraged to stand out.

One company we visited (an automotive assembly plant) had 10,000 employees and received an average of six suggestions per employee per month. This is a staggering number of suggestions—*each month.* Of course, not all of these ideas are Nobel Prize winners, but there are sure to be many worthwhile improvements there. Many of these ideas represent very minor changes that will make a specific activity just a little bit easier or faster or safer or less costly. Added together, the total value is immense.

We asked our host how many people were required to read, evaluate, and follow up on all of these ideas and the answer was "none." Each team handles the suggestions from its own members. After all, the team works as a unit to perform certain tasks, backing each other up and each knowing the entire process. It knows what is done and how and can easily assess the worth of each idea. Those that can be implemented easily and with little or no cost are put into practice. If engineering support, additional equipment, cooperation of another group, or approval is required, the suggestion is passed on as appropriate.

In addition to tapping into a vast reservoir of brainpower, the *Kaisen* approach establishes a culture of change. Employees at all levels of the company are constantly looking for ways to change, and improve, operations and activities and are therefore predisposed to not just accept, but to actually welcome change. Japanese employees are rewarded with bonuses (typically five months' pay) based on *company* performance, so all are keenly focused on whatever will make the company more successful.

Outside of such an environment, it is not uncommon to find varying levels of resistance to change. Change can be frightening, threatening, and uncomfortable. People react to impending change in many different ways. A major challenge for the implementation project team is to identify resistance to change in all its guises, determine the reason for the resistance or reluctance (there are many), and overcome it.

Resisting Change

A few years ago, I taught a class at the Righteous Rubber Company in the southeastern United States, and in attendance there was a gentleman I'll call George. Now, George was a senior manager in this company and had been employed there for over fifteen years. Throughout the class, George offered comments and discussions that all started out with the phrase "I really support this project" or "This is a really good (whatever) ...," or something similar, but there was always a "but" that followed. For example, "It's a really good idea to define the bill-of-material that way but that just won't work at Righteous because...."

Repeatedly during these discussions George would restate his support of the project and the ideas being presented but would always offer arguments as to why it wouldn't work. It became a game after a while, as I would respond patiently to each objection and George would just come up with another. The other attendees at the class would hide their embarrassment or roll their eyes during these seemingly endless discussions, and several came up to me during the coffee breaks and apologized.

George had an objection for everything. In his opinion, not a single change that was discussed would be possible in their environment. No argument and no amount of reasoning could shake him from this position. The interesting thing, for me, was the way George would first state full and unequivocal support for the project, then proceed to shoot down every single thing that was necessary to make the project succeed.

This is but one example of how people express resistance to change. There are many others. It's normal and natural to want to maintain the status quo. No matter how bad things might be in any particular environment, expect to find people who would rather keep things the way they are rather than to risk anything that is uncertain or at least unfamiliar.

The challenge for the implementation project is to do what they can to prevent, avoid, and overcome resistance to change, and to be able to identify its many manifestations so that it can be addressed.

Some resistance is no doubt the result of a perceived threat to basic human needs including security. Concerns about loss of employment, loss of position, loss of status, and diminished respect from others all come into play.

Threat of job loss can be easily handled. As recommended before (Chapter 3), a clear statement from the chief operating officer that no jobs will be lost as a result of this project should do the trick—but only if it is true. The first termination destroys the credibility of this promise. Was the system justified, in any part, on reduced head count? The word will spread. If labor savings are indeed a part of the plan, address the issue head on. Resolve (and tell the employees) that the net reductions will be handled through attrition and that people in positions that will be eliminated will be given other opportunities, many of which will be created by the new way of doing business (see Chapter 7).

As people are retrained and moved, be sure that the new position offers sufficient challenge and has sufficient status to be an acceptable substitute for the position that was lost. Be sure the employee understands the importance of the new duties and accepts it as a promotion. Also be sure to provide sufficient retraining to overcome any insecurity or fear of the unknown.

Another important factor is to tie success of the system (project) to the success of the company. Paint a mental picture of how the system will support the company's efforts to become and remain competitive in its markets (or reach new ones) ensuring continued employment for all.

Generally, people want to belong—make sure that every employee has an opportunity to become a part of the project and a contributor to its success. Even those who will have no direct contact with the system will be impacted by its use. Include *all* employees in your education and training plans and solicit their comments and assistance. There is less to fear for members of the implementation effort (with the entire company as an extended team) than there is for those merely standing by watching others make decisions that will affect them.

One effect of an integrated system is a much freer flow of information throughout the company. This, too, can be a cause of fear and resistance. Insecure people tend to hide information that they fear may be used against them. For an integrated system to work, information cannot be hoarded. Again, teamwork and education are the solutions. Encourage people to join the effort and become comfortable with the way things will be after implementation through an understanding of the goals of the effort—the way the system will provide benefits. Emphasize the integration, and how it multiplies the value of one's own data as they are combined with data from other areas. Be sure, too, that appropriately redefined measurements remove any risk that "someone else could get me in trouble."

Resistance also comes from those with a vested interest in the incumbent system, whatever it may be (including manual procedures as well as any computers or software). Occasionally, the authors of the old system will feel relieved to no longer have to support it, but usually there is a reluctance to let go of something created and nurtured. It is important to recognize the worth and accomplishments of the incumbent system and past efforts. Even if all agree that the current system is inadequate or has obvious flaws, there is undoubtedly some good there, and by recognizing it, you remove the tendency to defend the old system to the detriment of the new one.

The biggest source of resistance is perhaps the fear of the unknown, a contributing factor in most other causes of resistance as well. This is easily overcome with education and training. Help develop a mental picture in all employees of the new bright future that can be theirs. Show them how their daily tasks will be simplified, how their time will be put to more productive use, how their efforts will have more impact, and how the company will thrive as a result of their efforts. Tell them about the organizational changes that will take place (Chapter 7) and be sure to include a clear vision of how each will fit into the new scheme of things.

Fear of computers (or technology in general) can lead to fear of (personal) failure. Many very capable senior employees cannot type. Many very intelligent individuals feel totally lost in front of a computer screen of any kind. They are afraid of being embarrassed in front of their peers and subordinates. The important thing to do is to assure these people that their value to the company doesn't depend on typing skills. Second, provide hands-on training in a nonthreatening (non public) environment to help overcome unreasonable fears and demonstrate how easy it is to use today's systems. As the initial fear subsides, continue to emphasize the benefits of using the system:

the increased availability of information and improved management that results.

Don't ignore the fact that an implementation project requires extra effort on the part of all affected employees and rest assured that the employees will recognize that right away. They know they will have to learn new systems and procedures. There will be transition tasks like project team participation and data loading and validation. There may even be a period of parallel operation during which everything will have to be done twice (see Chapter 11). It is necessary to give each employee good reasons why this investment is worthwhile (see Chapter 10).

Finally, there are those who are just too comfortable with the way things are now to want to "rock the boat." This includes those who are nearing retirement and others who are simply disinterested in changing anything, for no particular reason. It might be hard to convince an employee with twenty-five or thirty years (or more) of service to the company, with only a few years left before retirement, that his investment in the implementation project will provide personal, long-lasting benefits. Let's face it, by the time the system is in and operating, he'll be gone.

In this case, you must engage the person's loyalty to the company (perhaps to guarantee his pension) and appeal to his ego—explain how his many years of experience with the products, processes, and operations of the company will be a valuable part of the implementation effort. The best part is that this is really the truth. Long-term employees can be of immeasurable help in dealing with organizational issues and identifying potential problems and solutions.

Signs of Resistance

Some resisters will announce themselves clearly, like our friend George at Righteous Rubber (as long as you read between the lines). Others will be more subtle.

One form of resistance is nonperformance. One who continually misses due dates for project deliverables is hurting the effort. Usually, there is a seemingly valid excuse. If others manage to complete task assignments, however, you have to question why this person cannot. When tasks are assigned, the assignee should agree to the amount of effort required and the projected schedule (Chapter 5).

Failure to execute, especially without notifying the team of the problem before the due date, is a serious breach of project requirements. If such a nonverbal resister is identified, he or she must be counseled and the problem resolved. It is likely that one of the preceding reasons for resistance exists; it (they) must be identified and resolved.

A member of the project team or a task team who continually misses meetings or sends a substitute is telling you that the project is less important than his other duties. Find out why he feels this way and emphasize the benefits of the new system (refer to the business case).

Someone who seems to be using the new system but continues to maintain card files, spreadsheets, or other "old" methods is not convinced that the system is "right" and certainly doesn't believe it's better. Letting go is difficult, especially if there is any doubt that the new system is at least as good as the old way. Apply liberal amounts of education and training to build confidence in the new system. Sometimes, the only way to wean users from the tried and true is to remove it (see Chapter 11) from consideration. It is much better if the employee voluntarily relinquishes old systems, but there are occasions where the issue must be forced.

Remember that there is safety in numbers. Reluctant individuals can be comforted and brought along through the magic of team interaction. A team offers mutual encouragement and support, and removes the feeling of isolation that can exacerbate insecurities and doubt.

Paving the Cowpaths

While I was in college, the school that I attended built a completely new campus out in the suburbs to which we all moved during my sophomore year. I lived in the dormitories that first year. They consisted of several buildings arranged around open "quadrangles," which is a fancy name for a square of grass. The architects decided to put sidewalks only around the edges of the quadrangles, leaving uninterrupted areas of lawn for everyone's visual pleasure. As you might have guessed, it wasn't very long before foot traffic had worn a series of paths across these open spaces.

The school's first response was to erect barriers at the ends of the paths with the expected "Keep off the grass" signs. Because the fences did not completely enclose the quads, only blocking the ends of the paths, the students naturally walked around the fences and wore a new set of foot paths into the turf.

The school finally gave in and paved the foot paths. Problem solved.

In the absence of established procedures (or in the presence of inadequate or unacceptable procedures), people will develop their own ways of doing things. Granted, they will most likely be undocumented and probably informal. Nonetheless, routines will be established that will accomplish the necessary tasks. At times, there may be official objection to the methods used, but this is usually a token effort (the result of an audit, perhaps) that is at best a temporary impediment to business as usual.

In most cases, if there is ever a move to document procedures, such as when preparing for ISO-9000 certification, these informal but established ways of doing things will become the formal, established *documented* ways of doing things, whether they are the best ways or not. The worn-in paths are paved and become officially sanctioned and permanent.

At the risk of stretching an analogy too far, let's say that the foot paths are not across a flat quadrangle of grass between dormitory buildings, but are in fact a series of winding trails through a hilly rural countryside. The paths are used by the peasants for years and years until finally a benevolent municipal government paves them (it must be an election year).

More years go by and the automobile makes its entrance. Eventually people walk less and drive more until another government initiative widens the paved paths into roads. Ah, progress.

Years go by and traffic increases. The road is no longer adequate to support the current volume of traffic but most people cannot even conceive of an alternative. They wait in line, honk their horns, and perhaps make derogatory comments about the other drivers' heritage, but they continue to follow the only roads they know because there's no alternative.

There's another election. A new administration takes over and promptly builds a highway through the area, but a special license is required to use the new road. New residents of the community immediately take to the highway as they can see it's the more direct and therefore fastest way to get from point A to point B. Most of the old-timers, however, continue to use the old road, despite its limitations and their earlier complaints, mostly out of habit and an underlying fear that they won't be able to pass the license exam for the new highway and, in fact, if they fail that exam, they fear that they might lose their driver's license altogether.

Eventually, the new administration threatens to close the old road. White-haired pickets march in front of city hall and grandmothers hold a sit-in at the entrance to the highway. Nothing changes.

Another election looms and the incumbent has decided to return to private life. Because he no longer cares about public opinion and is determined to leave the world a better place (in his vision) than when he took office, he orders the old road closed, the tarmac torn up, and a bed of flowers planted in its place. Only then is the new road fully accepted as the best way to get from point A to point B. Of course, some of the old-timers never use the new road (they just stop going to point B) and some simply move away.

The Union

A colleague of mine was a marketing support rep for IBM during the nineteen seventies, working out of the Detroit office. One of his accounts was a "big-three" automotive engine plant that was about to install an automated data collection system. In case you're not familiar with the term, data collection includes the placement of input devices on the plant floor that the workers use to report their activities, typically logging on to and off of each work order or job that they work on during the course of the day. Data collection systems gather the data and pass it to work order tracking systems and payroll programs.

Shortly before the system was to be installed, rumors began to spread. Before long, the union representative was in the General Manager's office threatening a strike. Plant workers viewed the data collection system as an invasion of their rights and privacy. They strongly resented having "Big Brother" in the plant, watching their every move. They were not about to let this happen.

After some explanations and serious negotiations, the data collection system was installed and, before long, was accepted by the workers. They had not understood (because nobody had told them) that the system was not intended to spy on them, would not infringe on their rights of privacy, would not be used to punish them. In fact, the data collection system made their lives easier by reducing the amount of manual reporting required. They no longer had to write down the job number, starting time, ending time, number of parts produced, and number scrapped. And they no longer had data entry people coming back to them the next day trying to decipher their handwriting and questioning what they had reported.

Things went along well for about a year, and then IBM came out with the next generation of data collection equipment. The company was very pleased with the progress made so far and decided to upgrade to the newer equipment. Once again plans were made, money was approved, and equipment put on order. Once again, the workers were not a part of the planning. And, once again, the rumor mill started spreading the word. This time, the union representation was in the G.M.'s office, strongly objecting to any plan to take away their precious data collection system.

This company made the same basic mistake twice. They failed, both times, to include the workers in the planning process and both times the workers objected to a change that they did not understand. Once they found out what was really changing and that the changes were to their advantage, they accepted these changes and made the new system a part of their everyday activities.

One of the most powerful and effective techniques for avoiding resistance to change is to not only inform, but also to *involve* the future users in the planning process. It is natural for people to take a proprietary interest in those things that they create themselves. By giving future users a part of the creation (or at least the specification) process, you can encourage this proprietary interest and begin to build ownership. Changes that fall down from the sky are scary. Nobody wants to be crushed under a falling rock. Get the users up on the mountain and let them help lever the rock onto the slope—and ride it down into the valley together.

Remember, too, that the people in the trenches—the users of the current systems and the future users of the new system—are in the best position to know what must be done and what makes sense. They can visualize more clearly how the changes will impact their daily activities, but only if they are educated about the new system early enough in the process to help avoid serious problems.

9. Education and Training

Many times, in these pages, I refer to the need for training and education as a cure for technophobia, as a way to build enthusiasm for the project, as the primary means of avoiding resistance to change, as the first duty of the project team (to educate themselves), and a major part of implementation planning, and more. This chapter presents some thoughts about planning an education program for a system implementation and some things to keep in mind as the system goes in and beyond.

First, let's discuss the difference between education and training. Education teaches theory, ideas, reasons, and the "why" of the system and the implementation. Training details the "how." Education is appropriate for all levels of people in the company from the Chief Executive Officer on down. Training is for the people who will be working directly with the system and its outputs. Training may include hands-on instruction and workshops working with the system itself or a sample of what the system will be when implemented (test database).

The most vivid analogy I've heard is this: think about whether you'd like to have your teenage daughter have sex education or sex training.

Education is not limited to theory and concepts. It can get fairly specific and might include demonstrations or some hands-on reinforcement exercises. And training might well (should, in fact) include some theory to provide perspective and context. So the dividing line is not crystal clear and not really that important. The point is that there is a range of kinds of training/education available and everyone in the company should receive some of it. The kind and amount will vary with the position, responsibility, involvement, and interest of the person.

Once again, I will talk about manufacturing system education, but the principles apply to any other kind of system.

Training or Education?

The difference between education and training is primarily one of depth and orientation. Training will focus on basic skills in a "how-to-do-it" orientation, whereas education will emphasize an understanding of the functions in more of a "why-do-we-do-it, and how-does-it-work" vein.

Training is aimed primarily at the actual day-to-day "users" of the system such as data entry operators, material handlers, accounting clerks, and production control personnel. It should provide enough understanding of the functions to instill a sense of confidence and provide a comfort level for the users without overwhelming them with the theory and complication of the system.

Training must provide the user with a familiarity with the day-to-day tasks that must be accomplished to keep the system functioning effectively. They should become familiar with the hardware (keyboard usage, system constraints, utility functions) and the software (menu organization, terminology, screen layouts, codes used) as well as the part that their function plays in the overall system utilization in the company. Hands-on instruction is very attractive in this environment.

Education, on the other hand, should be targeted primarily at the supervisory and managerial levels to provide insight and information that will allow the student to effectively apply the functions provided by the system to the management of the business enterprise. It is not enough (nor is it generally necessary) for the production manager to know how to go through the menus and enter a transaction. He or she must understand, however, what happens within the system that requires the transaction to be entered, and the effect that transaction has (when it is correct, incorrect, or missing) on his or her ability to control that part of the business.

For someone just getting started with a new system, it is all too easy to get sidetracked by the mechanics of the process and fail to obtain adequate management-level education. When the system first arrives, the most obvious and pressing job at hand is building the files. This is a rather mundane task in itself, and requires a working knowledge of the keyboard, screens, menus, and so on. Once the files are established, the users must be trained to maintain the data and produce the reports. The supervisors and managers are also new to the system, so there is a natural desire (and need) for them to also be familiar with the mechanics of the process ... after all, who are the operators going to ask when they have a problem?

Once the initial training is completed and the system is in regular use, however, the future success of the project often depends on how well *educated* the managers are. In order to advance the implementation beyond the basic functions (like handling inventory transactions and getting reports), the decision makers must fully understand the "why and how" and the interactions of the functional areas.

Management education must impart a wide-range view of how the various pieces of the system relate to each other. Detail is necessary, of course, but the detail here is very different from the detail given to the direct users, as discussed earlier. Managers need an understanding of concepts and relationships. They must be comfortable with the design of the functions and the reasons why applications work the way they do. It is from this kind of understanding that confident decision making and effective use of the system can take place.

Why Hands-Off Education?

There has been a great deal of emphasis in the marketplace on hands-on training, with many vendors offering more and more hands-on classes. Am I saying that hands-on training is bad for managers? Not at all. The problem is one of limited resources and relative focus. No system implementation project has an unlimited budget. No manager can afford to be unavailable while training for more time than is absolutely necessary. If scarce resources are to be applied to education, you must be sure that the investment will yield the best results possible. If time and money are of no concern, then by all means provide hands-on training as well as education for your managers. But if there must be a choice, then apply your resources where you will get the most effective results. Use the kind of education for your managers that will provide management skills and understanding.

When I speak of focus, my concern is that a manager might direct too much attention to the mechanics of the process to the exclusion of the "bigger picture": the management-level issues of the implementation. It is far too easy to get caught up with fighting alligators and forget that the original objective was to drain the swamp. Day-to-day tasks cannot be ignored, but they can and must be delegated. The key to delegation is to provide the end-user with sufficient high-quality training to allow her to function competently and effectively.

My concern is that education that includes hands-on experience with the system must do so at the expense of depth in the management issues. With a limited time to cover a topic, let's say a few days of education/training time for an application or subject, you cannot do it all and do it all well. Both education and training are necessary. The challenge is to apply the right kind of education and/or training to each person.

Three Levels

The types of education and training that are necessary to prepare all levels of the company to effectively use an MRP II system can be broken into three general categories: concepts, functional education, and user training. At the earliest stages, introduction to the concepts engendered in MRP II will provide the perspective that is needed to accept and understand the details provided in the other two levels. All employees should be introduced to concepts, appropriate to their level of involvement in the system. At the higher management levels, conceptual education might be quite extensive, a program in itself. For clerks and direct production employees, concepts might be presented in a few hours of meetings in the company cafeteria or may be included as an introduction to the detailed training materials. Concepts education for midlevel management falls somewhere in between.

Functional education is primarily for managers, at all levels, and will vary in depth and breadth from lower and middle managers (more detail, less broad in scope) to executives (less detail, all-encompassing). The point of this kind of education is to provide an understanding of the functions and features of the specific system being installed and how to use these facilities to support the decision-making process. Concept education provides context: an appreciation for how the pieces fit together (support each other) and the importance of each function to the overall operation and value of the system.

User training focuses on the day-to-day activities associated with system operation. It should provide those who will have direct contact with the system (data collection and input, routine use of system outputs) with enough understanding of the procedures and requirements to smoothly handle daily operations.

Format

There are a number of delivery methods in common use today. These include video tape, interactive video disk, computer-based interactive, live lecture (scheduled and on-site), seminars and workshops, hands-on training (both instructor-led and audio-tape/laser-disk/interactive-video–driven), and home-study courses. Some offerings include several of the preceding techniques. They vary in cost, duration, and appropriateness for the type of education or training desired. Some education programs will use several of these techniques in different combinations for the three target audiences (executives, managers, and system users).

Education in various forms is available from the software supplier and its affiliates and representatives, from professional organizations such as the American Production and Inventory Control Society (APICS, for general manufacturing management education), from consulting organizations, and from third-party suppliers.

A comprehensive education program for the entire company will most likely include a selection of formats and sources. Whatever the final configuration, the most important thing is to be sure that education is included in the budget and in the schedule. If there is an 80:20 rule for systems implementation, then the education and training activity represents a small portion of the overall implementation cost but has a major impact on the results.

For Executives

Executives must understand the impact that the system will have on the way the company does business. Typically, the highest-level executives will not be directly involved in the implementation except as strong supporters (see Chapter 3), but they can unintentionally torpedo the system if they don't understand how it fits. I could cite many examples of MRP II implementations where everything was in place and functioning well except that the owner/president/CEO did not really understand what a Master Schedule was or the impact of lead time (or lack of it). This can be very frustrating for everyone involved because all of the discipline and hard work that goes into maintaining the system can be destroyed in a heartbeat by an unaware executive.

One company comes immediately to mind.

A small electronics company in Pennsylvania called me in to help figure out why it was having difficulty shipping product on time. They had an integrated planning and management system in place and seemed to have good control of the major aspects of their business. There were no obvious mistakes or problems.

On further investigation, I learned that the owner of the company was an engineer who had developed the product that launched the company while working in his garage after hours. The company was now successful and growing, but the president was still working through the night, inventing new things. He had a habit of charging into the plant at any given time and saying, "Make me one of these." Nobody could say no to this man who signed all of their paychecks.

As a result, the production schedules were continually disrupted and customer orders were not completed on time. The company president was not being insensitive; he simply did not understand.

Concept (theory) education can be a program in itself and/or can be a part of more specific education and training programs. Generic (not system-specific) concept programs are readily available live or on video tape from industry consultants. Another source of concept education is professional societies like APICS, whose local and regional chapters sponsor classes and seminars in Inventory Control, MRP, Master Planning, Production Activity Control, Capacity Planning, and Just-In-Time, to coincide with the areas of testing for certification in production and inventory management (CPIM). An affiliate of APICS, the MGI Management Institute, also has a series of home-study courses (about $150 each) in similar topics for those who cannot or don't want to attend formal classes.

In addition to concept education, executives with operational responsibility will want some system-specific education. "Executive

Overview" classes are often available from the software supplier and sometimes from affiliated support companies, regional or local partners, or independent third parties who have experience with that brand of software. The larger the installed base for the package, the more likely there will be a choice of education suppliers.

A number of industry consulting companies, for example, "Big-6" accounting firms' consulting groups, can also present concept education on-site (at your facility), customized to your project's needs. Short programs on concepts are also offered at industry conferences.

For Managers

Functional education must be designed specifically for the system being used. At this level, it is necessary to use real examples and deal with specific features and procedures. The primary source of package-specific functional education is the software supplier and affiliated support companies or branch offices. For packages with a large installed base, there may be other sources available.

Functional education is at a level between concepts and hands-on training. A functional class in inventory management, for example, would discuss the specific organization and capabilities of the package, and will probably include sample screens and reports from the system. The class will explain how the function (or application) works, describe how it fits in with other functions and applications, and give advice on effective setup and use of the application. This last point includes not only features and functions of the software itself, but how these functions are applied to generate real business benefits. It's not enough to describe each inventory transaction (purchase receipts, issues to manufacturing, sales shipments, etc.). Functional education should put them in a business context, including the procedures and discipline that are necessary to ensure data integrity and timeliness, and how to use the system's functions to improve control and visibility of that area of the business.

There has been considerable discussion in trade magazines and elsewhere as to the relative merits of the hands-on versus the lecture-and-exercise approach to education at this level. Both approaches have merit and the customer must decide which is preferred.

In addition to scheduled classes, on-site programs (standard classes and customized offerings) are usually available from the listed sources. The advantage of an on-site class is that it can focus on the interests and concerns of an individual company, whereas a scheduled class must cover all areas to satisfy a diverse group of attendees. On the other hand, the ability to meet and discuss experiences with other companies is an advantage of scheduled programs.

Scheduled classes offer public enrollment at a fixed cost that averages several hundred dollars per student-day (1995 U.S. pricing). Costs for on-site classes typically run into thousands of dollars per day. Some education providers charge a fixed amount plus a modest per-student charge, whereas some are a fixed price with no per-student charge. Expect to pay for the instructor(s)'s expenses for travel and living for the duration of the class. On-site classes can be cost-effective. Typically, for the cost of sending four to six people away (considering class fee and travel costs for the attendees), you can educate as many as you can fit in a room. Most on-site class providers prefer a class size of thirty or less, although the circumstances and topic will influence the recommended class size.

One problem with an on-site class, however, is that it is more prone to interruption. Key people tend to get caught up in operational concerns and never come back from the coffee break. For this reason, it is wise to hold an on-site class at a nearby hotel or meeting center rather than in the company conference room. Another drawback of on-site education is the need to take a large number of key people away from their jobs at the same time. Sometimes, a split program (covering the same topics in separate morning and afternoon sessions) can be arranged to accommodate this need. Of course, this doubles the duration of the class.

User Training

User training must emphasize the "how-to" aspects of system operation but should not completely exclude the "why." User training must include hands-on exercises if at all possible. Computer-assisted training and laser disk or video training facilities are sometimes available to play a role in this segment of your education plan. Your software supplier might also provide some personalized hands-on training assistance at no additional cost as part of the installation support that they provide with your purchase of their system. Capable and knowledgeable members of your own project team can also become involved in the development and delivery of user-level training, especially with the help of train-the-trainer programs (sometimes called "T3" programs) from the software supplier.

Don't forget that users need perspective as well. Be sure that their introduction to the system includes enough concept and theory education so that they can fully understand their importance to the overall effort. They also must understand the "why" as well as the "how" to be properly motivated to maintain procedural discipline and to resolve questions or difficulties.

There are many illustrations of a clerk, worker, or other line employee making a simple error—doing something that makes no sense at all to someone who understands the system—that has devastating consequences and is a result of a lack of understanding on the part of the employee. It is usually not the employee's fault. Almost always, the root cause is that the employee was never educated about what they were being asked to do and what it means. Often, line employees are only shown how—"take this menu option, press this key, then enter the number here, then press this key ..." and not why. If there is any change or deviation from the norm, this person is completely at sea. He can either bluff his way through (guess at what to do) or seek help, which is not always appreciated, and help is sometimes not even available.

There is a risk of this happening with too much reliance on hands-on training, the focus of such training being strictly procedural and often ignoring context.

A Place to Play

It is helpful to provide a place for users to experiment with the system outside of the normal flow of business operations. Perhaps a small version of the "live" system with a representative subset of the database can be made available. Many systems allow the creation of a separate work space or "environment" that is separate from the live data where users can try new procedures, investigate cause and effect, and practice to gain confidence.

Many system selection and implementation projects now include a "conference room pilot" phase in which just such a sample system is set up to test applicability and develop procedures prior to implementation. The pilot system might be retained indefinitely for user training and experimentation.

There is a measure of fear and uncertainty in most of us that can be dispelled through the ability to become familiar with the system's operation, in an environment where we don't have to worry about interfering with company operations and where we cannot contaminate "live" data. We need a place to play, a place where we can make mistakes, and learn from them, without hurting the system or any of the other users while we learn.

Budgeting for Education/Training

There used to be a rule of thumb that the education/training budget for a system implementation should amount to 15% of the system hardware and software cost. With the price of systems (hardware in particular) coming down and the complexity going up, I'm not sure even that amount is adequate.

And yet, many companies budget far less than 15% and still expect to gain full benefits from the system.

It may sound strange, but the truth is that, no matter what is in the budget for education and training, in many cases some of this money goes unspent. This is especially true at higher levels—supervisors, managers, and executives.

The production control manager is scheduled for a week-long application class in how the system handles production scheduling and work flow management. Two days before the class is to start, a rush order comes in from your largest customer and the entire production schedule for the next five weeks has to be redone. Not only that, but now the plant is overcommitted and it's going to take some overtime, some shuffling of people and jobs, and constant effort and coordination to make sure all of the regular work gets out along with the rush order. Obviously, the production manager must cancel out of the class. He'll catch the next one.

When the next class comes around, there's another emergency or it happens during the manager's scheduled vacation or several other people are out or away and he cannot be spared. Eventually, everyone forgets that he never made it to class and by that time the system has been in for months and it's working fine without this key manager having attended system education.

There's always more to learn. Countless times, I have taught a class to experienced people and they have said that they learned new things, new techniques, and how to do things better, faster, or easier. In my own experience, as well, I never fail to learn something new even about systems I have worked with for years. I learn something new from every class I teach—each question or comment comes from a different background and perspective and triggers new and interesting ways to look at the same functions I know so well.

In any case, be sure to budget enough for education and training and be sure that the money is actually spent. Put someone in charge of monitoring the education plan. Keep records of every class scheduled and taken. What works best is a chart (on paper or computerized) for each person, specifying the classes needed, when scheduled, when actually taken, and comments from the attendee and the instructor (if available). Keep records of what types of education work best and which sources do the best job (this from attendee comments). Use this feedback when scheduling other people and other classes.

Is It Finished?

The job of educating the users is never done. People come and go, and they move around within the company. In addition, the company changes, the products and markets change, technology changes, and the system changes as well. Education provided at the outset will eventually be forgotten, will be overtaken by changes in the systems and the business, and will erode with personnel changes.

I had the opportunity to provide some education and training for a company in Baltimore a number of years ago, as they implemented a complete MRP II system. This implementation effort was well-organized, adequately funded, well-managed, and successful. After the bulk of the implementation work was completed, I gradually lost contact with this group.

One afternoon about three years later, I happened to be in my office (which at the time was in the Baltimore area) when I received a call from a senior manager from this company. He explained that they were having a serious problem with their planning system and asked for help.

I went right to the plant the next morning and they took me through an explanation of the problems and showed me a large volume of evidence and analysis that they had accumulated while trying to resolve the issue. Within a few hours, we were able to pinpoint the problem, which turned out to be a very minor procedural "glitch" that was easily corrected.

I was very surprised to hear from this company in this way because it had been such a good implementation with strong, dedicated people and a solid plan with adequate training and education. Their competence and success turned out to be their downfall. The project was so well received by the company's executives and senior management that the principals were all promoted and/or assigned to other important projects.

Unfortunately, the people that replaced those who had been rewarded with promotions and transfers were not given the benefit of the extensive education and training that the implementation team received. The specific problem that brought me back to this company was a peculiarity in the way it ran its planning system. There were good reasons for why it was set up this way, and it worked well in the early months. When the main planning manager moved up, he trained his replacement in the procedures as they had been defined, but did not provide enough background understanding. A slight change of policy in the purchasing department should have caused a change in the planning procedure to compensate. Because the planner didn't really understand why he did what he did, he couldn't anticipate the incompatibility of the new purchasing policy with the planning process. The result severely damaged the company's relationship with several key suppliers, caused much expediting and some costly production disruptions.

I'm reminded of a classic story that I heard a few years ago about a lumberjack who appeared at a lumber camp one day looking for work. When asked to demonstrate his skills, the foreman was amazed at the number of trees that the man could fell in an hour. He was hired on the spot and, in the next few days, proceeded to break every timbering record that existed.

The second week, however, his production fell off markedly. But, as he was still outproducing his peers, the foreman said nothing. The third week, he was beginning to fall behind some of the slower cutters, so the foreman decided to counsel the man. During the discussion, the foreman asked the lumberjack, "How often do you sharpen your ax?" The man replied: "Sharpen? Who has time to sharpen anything? I've been too busy cutting trees."

Do's and Don'ts

Whatever education strategy you develop, be sure to keep these things in mind:

- Don't assume that the users and managers can learn the system from the manuals. No matter how complete, comprehensive, or friendly the documentation is, it is no substitute for education and training. Just as the phone book has a great cast of characters but is weak on plot, the system documentation manuals have a wealth of information but are not organized to lead you through an explanation of how to use the functions.
- Don't put all your eggs in one basket. Be sure to spread the education around and include key people in all departments. Be sure that everyone has a backup. Make sure that every function can be performed by at least two people . . . just in case.
- Don't forget education and training in your budgeting and justification process. Include sufficient funds in all of your plans from the beginning (rule of thumb, 15–20% of the hardware/software costs). It is extremely difficult to go back to the budgeting folks after the fact and ask for training funds. Some vendors may try to lower their quotes to you by ignoring education costs—don't be fooled—no system is "turn-key," no system can be successful without educated users.
- Do include everyone. Even those who may not have direct contact with the system are still a part of the company and will impact the effort. The function of an integrated system like MRP II is to coordinate all activities to the accomplishment of your number one goal: delivery of your product to the customer on time. In the words of Eldridge Cleaver: "If you're not part of the solution, you're part of the problem."
- Do make your education program perpetual. As people move, are hired and leave, are promoted, and so on, and as your business changes through the years, it is necessary to reinforce the skills and extend the understanding to new circumstances and challenges. Once established, a level of competency is not permanent. It will deteriorate with time.

10. What's In It for Me?

Henry just wasn't sure. He'd heard a lot of talk about the new computer system that was being installed and how it would make everything easier and better. He had attended the orientation class and understood some of it . . . enough to know that things were sure going to be different.

The company president sent a letter to everyone, an open letter they called it, that was still on the bulletin board. "This is a critical project for the company," the letter said. "I'm counting on the full support and cooperation of every employee." "This new system is not intended to eliminate jobs." "No one would be terminated because of the new computer."

It said all that in black and white, right there on the bulletin board with the president's signature right there on the bottom. But, still, Henry wasn't convinced.

Henry didn't know anything about computers—not a thing. And he really didn't care to learn about them. After all, he was only five years away from retirement. Why go through all that with only a few years to go?

The president said nobody would get fired. Still, he had overheard some of the team members saying things like "Lead, Follow, or Get Out of the Way" and "We aren't going to let a few foot draggers hold up the project."

Henry wasn't about to lead anything with only five years left and not knowing anything about computers. He also didn't feel much like following. Couldn't they just leave him alone? What exactly did they mean by "Get Out of the Way"? Did that make him a foot dragger?

The truth of the matter was that Henry was afraid and embarrassed. Or, more correctly, afraid to be embarrassed. After more than thirty years with the company, he had built up a measure of status and respect with the other people in his department, most of whom were quite a bit younger. They looked up to Henry for his experience, often asking how to do this or what to do about that. But this computer thing?

He could picture it clearly. They'd sit him down in front of that thing and tell him to do something and he wouldn't have a clue. He knew he couldn't type. He didn't understand half of what they were talking about when they did their orientation in the department. If they asked him to do something on that computer—anything—he'd look like a fool. He'd lose all of the respect he had worked so hard to build up over thirty years. But it was surely a trap, wasn't it? He needed another five years to retire and they were going to cheat him out of his pension. "Get Out of the Way?" Gladly, but how could he do that and still keep his job?

Computers are undoubtedly a part of every life today in everything from the cars we drive to kitchen appliances to the checkout counters in stores, yet there is still a great deal of computer phobia and general misunderstanding about what they are and what they can be expected to do.

In part, it's a generational problem. As I write this, I am in my mid-forties but I have worked with computers for years. And I have an engineering background. As I look at other people who are my age or older, with the exception of technically oriented folks like myself, there is very little understanding of computers and considerable fear and distrust.

Younger people, especially those in their twenties today, have been around computers much more and are less intimidated. Most had some exposure to computers in school and they are far more used to living in a technical world.

Is it any wonder that a company's senior managers and most experienced employees are often the last to embrace computer technology? As my generation moves on and the younger one moves into senior management positions, the adoption and use of technology will accelerate. Resistance to change that is based on technophobia will lessen.

On the other hand, sometimes there is an inflated view of what technology can do and this can be as dangerous as fear of technology. This issue is addressed in Chapter 13.

Please don't think that I am condemning an entire generation or unfairly categorizing everyone by age group. I'm not saying that real change must wait for the older generation to retire or die. My only point here is that the definition of what is "new" and "unfamiliar" is changing. We are fighting a considerable measure of computer phobia today that will decrease as time goes on.

At a company installing an MRP II system, there was a key supervisor, let's call him Fred, who was uncharacteristically uncooperative regarding the system project. In all other phases of his work, Fred was a model employee: hard-working, loyal, reliable, and dedicated. When it came to the new computer system, however, he just wouldn't go along. He was not openly hostile, but neither was he supportive. Whenever he was assigned a task, it was either late, or poorly done—very unusual for Fred. When questioned about this poor performance, there were always legitimate-sounding reasons for his nonperformance, but at the same time, the project managers could see that the tasks could have been completed and could have been done right, if only Henry had wanted that to happen.

This was a real problem because Henry was such a good employee in all other aspects of his job. He was also highly respected and any punitive action would demoralize many others and cause even greater problems.

The situation was rapidly reaching the point where something had to be done. Henry's subtle reluctance was becoming obvious to everyone, and if it was not addressed and resolved soon, the project would begin to wither and die. Would they have to fire him?

Finally, one of the software supplier's technical support people built up enough rapport with Henry to get him to open up over a beer at a neighborhood watering hole.

"I'm not going near that computer, no matter what. I don't care if they fire me, I'm not doing it."

"Oh, come on. It's not going to bite you. And you can't hurt it. There's nothing you can do at that keyboard that will hurt the system in any way. It's really no big deal."

"For you, maybe. You're a young guy and you know all about computers."

"And?"

"And I've never touched one of those things in my life. And I ain't about to, either."

Later in the conversation ...

"They'll laugh at me."

"Who will?"

"Everybody. Maybe not out loud, but they'll be laughing inside. I know I'll look stupid and they're gonna be rolling their eyes and wondering how an idiot like me could have fooled everyone for so long."

And that was the real problem. This poor guy was terrified that thirty years worth of work meant nothing because he couldn't type, and he thought he would look the fool when he sat down at the keyboard.

Some *very* private lessons were arranged. After hours and off-site, Fred received some basic training in keyboard operation, familiarization with how screens were laid out in this system, signing on and off, and other basics. Before long, he was in regular classes with the other employees, learning the specifics of putting the system into operation. He became a real driving force in the implementation project.

Although few will express it verbally or quite so bluntly, everyone wants to know how changes will affect them personally—the "What's in it for me?" that is circling our bewildered friend here in Latin (see Figure 10-1). It is critically important to answer this question with an answer that motivates support and participation in the project. There can be no disinterested observers, no innocent bystanders, no "wait-and-see" skeptics.

The project leader, project team, and company executives must create a vision in each employee's mind of a future that is attractive enough and important enough to make each employee want it to come into being. All will be asked to make some extra efforts during the implementation. All will be asked to relinquish long-held comfortable and trusted processes, procedures, reports, spreadsheets, forms, filing systems, methods, techniques, relationships, and informal information sources in favor of something entirely new, unknown, and (in their experience) unproved. That's asking a lot.

Why should they go through all of this? Why would someone want to risk massive change and give up those things that she knows will work. Why should he put in the extra hours and go through the

Figure 10-1. "What's In It for Me?"

"growing pains" of implementation?

The only answer is because they see that there is sufficient reward to justify the cost. Sufficient *personal* reward. Personal reward doesn't mean money in their pockets. I'm certainly not suggesting that you bribe every employee to go along with the changes. (That's an interesting idea, don't you think? Just kidding!)

Personal reward can take many forms and since we are talking about business systems, we should be thinking in terms of working conditions, opportunity for advancement, reduction or elimination of onerous tasks, increased effectiveness, more time for more challenging and important activities, and better prospects for continued employment if the company is floundering and the project can help it remain in business.

Each person has to feel that it is in his or her own personal best interest to bring the project to successful completion. Each person will have his or her own "hot buttons," things that motivate them. Fortunately, most of the motivators will relate to the overall success of the company; the challenge is to bring them to the fore so that the employee is fully aware of how his own best interest is also in line with the project and the company goals.

Obviously, if the employee fears for his job, it will be very difficult to get him to concentrate on what's good for the project or the company. As discussed in Chapter 8, basic human needs must be dealt with first, that is, be sure that such things as security, belonging, and self-esteem are satisfied before looking for higher-level motivators.

Winning Attitudes

8:00 A.M. The rest of the project team will be drifting in over the next half-hour. Right now, it's just the project manager and I, being philosophical over the morning's caffeine.

"It's really interesting," I say, "to see how people are reacting to the new structure." This company has just been "cut loose" from a large organization and is now an independent entity rather than merely one of many divisions in a Fortune 100 behemoth. They are feeling the real sting of open competition for the first time. Some fold under the pressure, others thrive. All are more than a little scared.

The project manager chuckles. Actually, it's more of a harumph. I know what he means. From my perspective, looking in from the outside, I can afford to be amused. He's in the middle of the battle. Although it may not be literally life and death, it certainly is a life-and-death situation for the company and for hundreds of jobs.

Many employees from this division took advantage of early retirement incentives that were offered as a part of the restructuring. Those who chose to stay, or didn't have enough time in to have the option, have been thrust into a new world. The current vocabulary would call it a paradigm shift. The old priorities are gone. As a division of a large multinational, this group did not face the realities of the market in the same way that an independent company would. They were fairly well shielded from the harsh realities of profit and loss and unbridled competition. In this new incarnation, they are just one of many competitors in a very dynamic market. The new order is all about being first and being the best. They are now in the "real world" and it's a bit overwhelming.

"The attitudes on this team are terrific," I say. They are evaluating a packaged, midrange-based MRP II system and comparing it to the custom, mainframe-based system that will no longer be available to them under the new structure. "When you compare the function of a package to what's in custom code, there are going to be differences. It would be easy to complain and find excuses. This team has looked with an open mind and has found much to be pleased and excited about."

"They're good people," the P.M. says. I agree. "They know what they need and they also know that the mainframe has limitations, too." Like once-per-month MRP runs.... On somebody else's schedule. Like horrendous corporate charges for service. Like a system that hasn't changed in ten years and is unlikely to change much despite advances in management theory and technology.

"If we can't get it together—and fast—we won't be around a year from now. That's fairly good incentive for being adaptable. We're in no position to be sitting around singing 'Auld Lang Syne.' We have one chance and one chance only."

"In any case," I say, "the team is reacting positively. I've been to companies where you can feel the tension as soon as you walk through the door. 'Cut it with a knife,' as they say. Pressure exists. Every company faces competition. In some of them, the reaction is panic and blame laying. In others, the team rallies and it drives them to excel—focuses their energy into positive results. That's what I see here."

The P.M. smiles, justifiably proud of his team. "It had to happen. If we want to stay in this business, we have to get tough. Everybody knows that it should have happened five years ago."

Indeed they do. Every analyst, stock picker, and newspaper financial columnist has been saying the same thing for years. In a large company with a significant corporate bureaucracy, the incentives get lost in company politics and hierarchy. Large companies cannot move fast enough to react to market changes. An elephant can't keep up in a *pas-de-deux* with a cheetah. Thus, we have the deconglomeration of America. The Bell companies were forced to break up by the courts and it has turned out to be the best thing that could have happened to them. IBM is now beginning to understand this concept and the initial results are very encouraging. Others will follow.

"This is a good package," the P.M. says, "and the fact that it will

grow and change with the technology is especially exciting. But tell me one thing: Is this the best package on the market?"

Now it's my turn to smile. "It's as good as any and better than most," I reply. "There are more than two hundred packaged MRP II solutions out there, and they are far more alike than different. What makes a company successful with MRP II is the team that implements it, getting user acceptance (through education), and having a strong, visible management commitment—all having nothing to do with the software. If the software works, and just about all of the systems do, what you need most from the vendor is support. Do they keep the package up to date? Do they respond to your questions and problems? Will they be there five years from now? These are the critical questions. Don't worry too much about functionality. It's there."

The P.M. nods, looking perhaps a little relieved. "We're in good shape, then," he says. "If those guys would just get their butts in here we could get on with it."

As if on cue, the door bursts open and a half dozen of the sharpest, most enthusiastic people you'd ever want to meet come walking in, ready to meet the challenge.[1]

Everyone has an ego—some more visible than others. It helps to recognize what makes people feel good about themselves and their accomplishments and appeal to that part of their character. Most people want to do a good job and want to contribute to the well-being and success of the company. In introducing the system and its functions, relate the upcoming changes to how they will enhance each person's contribution to the business and how the system will help the company become and remain more competitive and more successful. Each task and each feature can and should be related to positive results in this way. Help each employee build an association with the success of the implementation, the success of the company, and personal success.

Be generous with titles. It costs nothing to give someone the title of task leader, department coordinator, project/subproject/task/department/function manager/coordinator/leader/liaison/commander, or tzar. If nothing else, titles recognize the importance of what is being done in a highly visible way.

---

[1] Adapted from my July 1993 column in *3X/400 Systems Management*.

Visibly recognize accomplishments. Celebrate the completion of each task, phase, and major milestone. Budget a few dollars for cake, pizza, and other treats for these celebrations. Publish the good news in the company newsletter, if there is one, and if there isn't, create a newsletter specifically to keep everyone up to date on project progress. Be sure to give credit where credit is due. Name all of the participants and contributors to each success. Send thank you notes and/or memos. Have certificates made that can be presented to those who help the project and as recognition of education and training completed.

Continually sell the merits of the new system and new procedures. Communicate in every way possible the expected business benefits. Have each department prepare a departmental impact statement—a subset of the business case. In so doing, they will be forced to examine the direct impact of the upcoming changes and will justify their own efforts in a way that relates more directly to their daily tasks and challenges. The results of this exercise will also provide some insight as to whether the department understands its role and requirements.

Make sure everyone understands the impact of not completing a project on time. Do this carefully so that it is not threatening to the individual and does not add to any existing insecurity. The fact of the matter is that successful system implementation can be a critical factor in the future competitiveness of the company. Make sure that this is understood. The success of the company means the continuation of employment and more opportunity for advancement for individuals.

Publish your measurements. The business case and project plan will include a number of measurements that will be monitored throughout the implementation process (and beyond). Post major measurements in a conspicuous place and keep them up to date. Have simple explanations of the meaning (impact) of each and track progress toward the goals and milestones set forth in the plan. Be sure that individuals understand their contributions toward meeting these goals and encourage them to take a personal interest in meeting or beating the targets.

Although interdepartmental rivalry can adversely affect the smooth operation of the company and can get in the way of the implementation of an integrated system, a little healthy rivalry within the implementation project can be beneficial. Encourage task teams to exceed expectations, and reward their success when they do. Recognize the team that does the most or the best as task team of the week (or month) or by a measurement that can be applied to any task, such as a 110% club (finished the task 10% ahead of schedule or loaded 110% of the expected data in a given time). Attach a reward to the award—pizza party, free coffee for a week, T-shirts or baseball caps, picture in the company newsletter, and so on. When one team sees another one rewarded in this way, it will be encouraged to do even better (and be rewarded even more).

Motivators

- Create a vision to which each employee can relate.
- Instill in each person personal reasons for supporting the project.
- Find out what each person's "hot buttons" are and appeal to them.
- Celebrate, publish, recognize, and reward.
- Be sure that basic human needs are satisfied before appealing to higher motivations.
- Encourage excellence by rewarding task teams for accomplishments, encouraging intertask team competition.

11. Cutover (Letting Go)

In 1981, I had my first consulting experience on a systems project for a manufacturing company. I could barely spell MRP at the time but, nevertheless, I helped this company through the usual steps of requirements analysis, preparation of the request for proposals, and system selection. Once the successful vendor had been selected, the company agreed to continue working with me through the implementation process.

This company had been strictly a research outfit, designing and prototyping new items for various military organizations. At the time I was involved, they had just received a contract to begin actual production of one of their designs in some quantity. Real manufacturing was something quite new to this company, thus the decision to install a new system to support this new set of requirements.

In their prototyping work, the company had a modest amount of inventory to keep track of and some very basic bill-of-material data. Because most of the employees were engineers, the company had one of them write some programs for an old Data General computer that wasn't being used, to hold the inventory and bill information.

Quite early in the new system implementation process, we had the inventory tracking and bill-of-material applications ready to go. One of the engineers helped us convert the data from the Data General (nicknamed "General Lee") to the new system, we trained the users, and we started using the new system. The trouble was that the engineers continued to use the old Data General while putting the same data and transactions into the new system.

We finally had a steering committee meeting at which we agreed that the new system would not become the main source of information as long as old General Lee was still around. So we planned a funeral.

One of the data entry clerks had a sheet cake decorated with a tombstone saying, "R.I.P. General Lee" and we held a ceremony at which we unplugged the old machine and hung a wreath over its cabinet. From that day on, the new system was accepted and used.

At another company, production scheduling was done using index cards attached to large boards that lined the walls of a "scheduling room." After the equivalent production scheduling and tracking functions were implemented on their new system, people continued to keep the schedule boards updated and workers would invariably refer to the boards when they needed the schedule information rather than using the terminals provided to look into the new system.

On a Saturday morning when nobody was around, the implementation project leader, production manager, and several other project team members disassembled the scheduling room, burned all the cards, and hauled the scheduling boards to a remote warehouse location. On Monday morning, there was a lot of complaining and a small measure of panic, but by the end of the week, the new system was being fully utilized for scheduling and tracking production.

Just last month, while discussing an upcoming implementation process with a client, the company's Chief Financial Officer asked me how long I thought the company should plan to "run parallel" with the new and old systems. My answer was no time at all. I don't believe in parallel operation.

This is a controversial position, especially with financial executives. The traditional approach (and financial people tend to be more traditional than most) is to start using the new system while maintaining the old one for some period of time. During the period of dual or parallel operation, the results will be compared to verify that the new system is consistent with the incumbent one before trusting the new system to do the job. Typically, parallel operation must proceed through at least one cycle including a "close," which in the financial world is either one accounting period (month) or several months (at least) and one year-end cycle.

On the surface, parallel operation seems like a very sensible idea. After all, the new system will be handling vital company information. You want to be sure that the new system works, so you can trust it.

You also want to hold on to what you know works (the old system, whether manual or automated) until you are sure that the replacement is adequate.

The problem is that nobody has the time or resources to do everything twice. Inevitably, one or the other (or both) system(s) will not receive the full attention that it needs. If the objective is to compare the systems, it won't work. If one or the other (or both) is/are not being fully maintained, any comparison will be severely flawed.

In all likelihood, it is the new system that will suffer from lack of attention. The company is presumably still operating from the old system (until the new one is proven), therefore, the old system will get priority. In addition, the users are still not completely familiar with the new system, so they are apt to make errors, skip steps, fall behind in procedures, or get frustrated with it. So, how can parallel operation be a fair test of the new system's capabilities?

Even assuming that both systems can be fully maintained during parallel operations, there is the additional task of comparing results to verify correct operation. It is likely that files, data structures, and output formats of the two systems will be somewhat different (if the new system is exactly like the old one, why change?), so comparison may not be as simple as you might think.

Given all of that, how can you be sure that the new system works correctly in your environment and with your data, and that nothing will be lost in the transition? That is a challenge that must be faced by any organization that is changing systems. Since parallel operation is impractical, what else is there?

The Conference Room Pilot

System implementers have developed the Conference Room Pilot (CRP) as a method for validating a new system before cutover. The CRP, if effectively executed, also offers an opportunity for training users and for developing at least the basics of a set of new operating procedures at the same time. Here's how it works.

The new system hardware and software are installed, complete enough to run all functions to be implemented. The pilot test system does not need to be large enough to support the entire company. In other words, a smaller version of the hardware and a subset of the database can be used.

This process is called a *Conference Room* Pilot because that is where it normally takes place. A number of terminals are set up in a room where the teams will gather to run test scenarios.

During the pilot test, all functions will be simulated, using real data, by some of the actual future users of the system, but the simulation will not necessarily be in "real time," that is, the time relationships between tasks and functions are not important. Each process will be run, from start to finish.

As an example, let's say Purchasing is included in the new system. A team will be assembled to pilot test the purchasing process. The entire process will be exercised, starting with identification of a need, then communicating that need to the buyer, identifying potential suppliers, quoting or negotiating, creating a contract (if appropriate), requisition, approval, purchase order release, receipt of acknowledgment, receipt of goods, incoming inspection (with rejects, return to vendor, quality hold, etc.), move to stock, notification to the requisitioner that the items have arrived, invoice matching and validation, invoice transfer to accounts payable, release for payment, payment, close (P.O. and invoice) and archive to history, vendor performance analysis, and so on.

As each step is performed (by the future user), any missing function, inconvenience, or other difficulty is documented and the team resolves how the problem will be handled. Procedures are drafted (rough form) that will be the basis for the new procedural documentation after the system is implemented.

The Conference Room Pilot is an isolated test case, but it does use real company data and real users to validate the adequacy of the new solution. These users get a good taste of what the process will be like after cutover (training) and problems should be uncovered through this process.

It is important to test all variations and combinations in the pilot scenarios. As each isolated area is exercised, interactions with other scenarios can be checked—inventory handling processes can pick up information from the purchasing receipt scenarios or vice versa.

This part of the CRP does not test system speed (responsiveness) or identify system-level procedural conflicts (usually, there are a number of other functions that cannot be run while MRP calculations are being done, for example). The software supplier and the IS department should address these issues separately from the CRP.

No matter how much testing is done in the conference room, some people will not feel confident until the system is tested under true operating conditions. That is understandable, but, again, parallel operation doesn't work. After sufficient pilot testing to validate system operation and complete initial procedure development, there will still be some uncertainty and some room for problems when the system actually goes "live." Unfortunately, this is unavoidable, but it is manageable. The risk can be minimized by using a phased cutover to turn on portions of the system over a span of time.

Phased Cutover

It's a scary thought—one day your company is operating with the old system and the next day it is gone and a new one (not yet completely tested and proven, and one you are not yet comfortable with) has replaced it. Seems like a prescription for disaster, doesn't it?

Yes, it can be a disaster, but, with proper preparation, it can be the most efficient and effective approach. There is still a practical problem of changing *everything* all at once—often there is just too much of a change to manage all at one time.

The compromise is a phased implementation, as alluded to in Chapter 5. In that chapter, I laid out a sample implementation sequence at a high level (applications):

	Primary Responsible
PHASE I: Customer Service	
Inventory (finished goods)	Materials
Order Entry	Customer Service
Accounts Receivable	Accounting
General Ledger	Accounting
PHASE II: Materials	
Inventory (raw materials)	Materials
Purchasing	Purchasing
Accounts Payable	Accounting
Bills-of-materials	Engineering
Material Requirements Planning	New function
PHASE III: Production	
Work Centers and Routings	Engineering
Production Control	Production
Bar-Code Data Collection	Production/IS

Engineering Change Control	Engineering/ Production
Capacity Requirements Planning	New (production)
PHASE IV: Planning	
Forecasting	Marketing
Master Production Scheduling	New function

In this list, there are four groups or phases, each with several applications being installed at one time. The applications in each group were chosen because they relate to each other and to a business functional area, and therefore should be turned on at the same time. And different people are responsible for the different applications in a group, and some people are involved in multiple phases, so that the work load is spread around. Each phase would have a single cutover date.

In conjunction with application phasing, or in place of it, an implementation can also be phased by department or product line. If it is a large company or one with separate and distinct business areas, each area can be brought up on its own schedule, coordinated with the overall (corporate) project plan. One company I know, for example, has a fabrication area where raw metal (bars, sheets, rods) is machined into parts that are transferred to the assembly area and also sold to outside customers. Another part of the plant assembles pumps and valves, which are used internally and also sold outside. A third business area uses both fabricated parts and pump assemblies to make a consumer product.

Because each business area is operated fairly autonomously, each could be brought up on the new system on a separate schedule. In this case, it was fabrication first (simplest process and lowest parts count), then pump/valve assembly, and finally product assembly.

The challenge in this kind of phased cutover is to coordinate the old and new systems where the business areas interact. When the fabrication area is on the new system and the rest of the plant is still on the old one, internal requirements for fabricated parts must be gathered from the old system and passed or entered into the new one. Inventory records must be coordinated for common part numbers. Separate planning (MRP) runs are required in each system, whereas a single planning run would have previously covered the entire operation. Completed parts must be transferred from the new system inventory to the old system inventory when "sold" from the fabrication plant to the assembly plants. And so on. In addition, costing and financial data must be transferred from whichever system does not include the operating financial applications at the time.

As you can see, it is not easy to operate in a split environment. Splitting systems should be avoided if possible, but if it must be done, make the time period as short as possible to reduce the amount of extra work required to keep things in synchronization.

If applications must be split in a phase-in process, there are hard ways to do it and less hard ways to do it. Look at the interfaces between applications. The more data that are exchanged between applications, the more difficult the temporary interfacing will be.

The easiest split is probably between operational applications and financial modules. Even though most systems will generate journal entries to the general ledger for many types of activities on the operations side, these transactions are usually easy to "trap" and/or "bridge" over. The most difficult aspect of this split will be the relationships between Purchasing and Accounts Payable and between customer ordering/invoicing and Accounts Receivable, which are likely to be more tightly integrated.

Trying to split or phase within the operational applications is very difficult. Having inventory on one system and production on another, for example, would present a very complex interfacing problem. The phase-in list shown earlier makes the most sense for a new implementation, as compared to one in which an existing system is already providing the listed functions. Where the new system is a replacement for a functioning incumbent system, an all-at-once cutover provides the simplest path from a systems perspective. If this is impractical, thoroughly analyze and plan for the necessary data exchange between old and new and minimize the time that you will be required to make this exchange. The same considerations apply to other phase-in approaches such as bringing up one business unit or product line at a time.

Data Conversion

Another aspect of cutover is when and how to convert the data from the old system to the new one. This discussion will assume that the entire system is being brought up at once, but the same considerations apply in a phased approach.

There are basically three kinds of data resident in a system:

- Master Files: Contain relatively static data that define and control. Examples are an Item Master File, General Ledger Chart of Accounts, Vendor File, Payroll Tax Table, and so on.
- Transaction-Driven Files: Contain activity-based information such as purchase orders, customer orders, production work orders, invoices, journal entries, and so on.
- Archives: Historical records typically copied from transaction-driven files and saved either on-line or off-line on tape.

Generally, Master Files must be converted all at once. Master File conversion can be distributed by application phasing, but all of the Master Files used by the customer service applications, for example, must be converted at once.

If the phase-in strategy causes multiple users of a single file to be in both systems at the same time, duplicate files must be maintained. Let's say that Purchasing is in phase 1 and Accounts Payable is in phase 2. As phase 1 goes live, the vendor master would be converted into the new system, but it continues to reside in the old system to support A/P. Any changes to the vendor data (change of address or phone number, new vendor, deleting an obsolete record) must be made to both files. Some master files are also used to accumulate historical data or statistics. By using the Vendor File example, it may contain the total year-to-date purchases, discounts taken and lost, vendor performance statistics, date of last purchase or payment, and more. If this is the case, when A/P is converted with phase 2, the statistical data that are tied to A/P activities must be updated in the new system's Vendor File without corrupting any statistics that are tied to Purchasing activity.

Archives are generally not a problem. When the application that uses the archived files is converted, either convert the archived data or start fresh with a new archive on the new system. Conversion is preferable, of course, so that you will still have access to all the old history. Because systems are all (internally) structured differently, there will be a file conversion effort required (write a program to translate the data from the old system's format to the new system's requirements). Sometimes the new system vendor will assist in this conversion, either as part of the entire effort or at extra cost.

Transaction-driven files offer some interesting considerations. All-at-once conversion is certainly feasible, but may not be the most practical approach. Because custom file conversion programs would be required to electronically transfer the data, this can be an expensive proposition. Manually loading thousands of orders or invoices is also unattractive.

What I recommend in most implementations is a limited-duration transition plan for each of these transaction-driven files:

- Begin operations in the new system at initial cutover.
- Continue processing in the old system those activities that existed at the time of cutover.
- After a brief interval of dual operation, manually convert the remaining "old" activity records to the new system.
- Convert archives and historical data from the old system to the new one.

This approach avoids laborious or expensive data conversion but causes some confusion and extra work during the transition period—which is why I emphasize that the transition period must be of limited duration.

Let's say there's a purchase order file in the old system at the time of cutover that contains two thousand active P.O.s. Let's also assume that the typical lead time for the majority of these P.O.s is two to three weeks. All new P.O.s will be entered into the new system after the cutover date. Receipts, invoices, and so on, for "old" P.O.s are processed in the old system for no more than one month. At the end of the month, there may be only a few hundred P.O.s remaining in the old system. These are then manually loaded into the new system and the old system P.O. activity is terminated.

Cutover

- Parallel operation doesn't work.
- The Conference Room Pilot is a recommended method of proving applications, training users, and developing procedures.
- The phase-in of application groups is a typical approach for new system implementations.
- The phase-in by business unit or product line can sometimes be a practical alternative.
- Data conversion recommendations (from an existing system) vary with the type of data.

- Phased implementation greatly complicates conversion and interfacing requirements.
- A limited-duration transition plan is recommended for transaction-driven data.

12. Measurements

There's a consumer goods company that I worked with a few years ago that had an inventory problem. The amount of (finished goods) inventory was much too high and increasing. Now, it turns out that this company had a measurement problem that was a major underlying cause of the high inventory. I had the opportunity to work very closely with the Master Scheduler at this company—the individual who managed the overall production plan. Master scheduling, in a nut shell, is the process of matching supply and demand. Demand comes in as both forecasts and customer orders, supply is the production capability (and all supporting activities like purchasing and materials management), and the trick is to come up with a master schedule (production plan) that best utilizes the company's resources (supply) to meet demand.

This company was using some fairly sophisticated computerized tools to help develop an optimized master schedule—putting the resources to best use while meeting company objectives in terms of satisfying demand and making a profit. Jim, the Master Scheduler, would use these tools and come up with a fine, balanced schedule. Then, as a last step, he would total up the number of units to be produced each month. If the total was under 2.6 million units, then he would manually adjust the schedule until the total was 2.6 million. This adjusted plan was then released.

Why 2.6 million units? That's an excellent question. It seems that this company had done a study that took the total production quantity for a year and divided that number into the manufacturing budget to determine a cost per unit. You can argue with the fact that all units were considered equal in this analysis even though they ranged from one-ounce samples to fifty-five-gallon drums (the company packaged liquid chemicals), but let's not quibble. This was an accounting exercise and the number was used in determining profitability of products and for preparing quotations for new business.

Unfortunately, the cost-per-unit number became somewhat of a magic piece of "truth" that escaped its original purpose and began to be applied to other things. Because this number represented a way to set and evaluate the manufacturing budget, the logic went, then the quantity used in its calculation must be the quantity required to make it valid. The annual production quantity happened to be a little over 31 million units, therefore, 2.6 million units per month was the goal.

One month's master schedule that I saw totaled just over two million units. Jim used his experience and judgment to add a few hundred thousand units here and a few there to bring it up to 2.6 million. You will not be surprised to hear that, at the end of the month, there were about a half-million more units of finished goods inventory in stock than there were at the beginning of the month.

Jim understood the problem, but he was strongly motivated to come up with a schedule for 2.6 million units each month. Anything less was simply unacceptable and would be returned to be redone. As a footnote to this story, Jim was eventually fired because his performance was deemed "unacceptable" by the company president, partly because Jim insisted on trying to point out the fallacies of scheduling 2.6 million units just to fulfill the accounting cost-per-unit prophesy.

This company invested heavily in master scheduling tools but did not allow them to function. It had great difficulty controlling inventory growth and never realized that it was an accounting measurement, not even an inventory or planning/scheduling measurement that was the direct cause. Nevertheless, it was a measurement problem. The master schedule should have been measured against company goals like customer satisfaction (order fill rate, for example), or profit or margin targets.

Different Problem, Same Cause

At the same time, at the same company, the production manager was having problems producing those 2.6 million units. In an attempt to improve "efficiency," the manpower budget for manufacturing was reduced while production schedules were held at the same level. Because efficiency measures output per hour of direct labor, reducing the labor and keeping the output constant will obviously increase efficiency.

Bob, the production manager, was constantly driven to do more with less. Equipment could not be properly set up and operated, some lines went idle because manpower was not available, any minor "glitch" in the process became a major disaster, and the poor guy was aging right before our eyes.

The efficiency goal was driven by the same cost-per-unit focus. Serious consequences of undermanning were ignored in pursuit of this one objective. As a result, manufacturing was forced to give preference to products with high-quantity-per-hour run rates at the expense of anything that had a higher labor content or slower run rate. The plant was always overscheduled, so there were always a selection of jobs to choose from and the driving factor was the efficiency measurement. Thus, important work (products with shortages) was deferred whereas product that was not needed (had been added to the master schedule to meet the units-per-month target) was produced and moved into finished goods inventory. Inventory increased while customer service languished at an undesirably low level.

The sad part of this situation is that the direct labor content of this company's products was about 6% of the cost of goods. If all of this effort and agony managed to reduce the direct labor content by 10%, the net effect on the company's profit would be a paltry 0.6%.

How could this company get so focused on such a (relatively) insignificant cost? In part, it is the fault of the measurements, in this case, the accounting system. Traditional cost accounting evolved in the last century when labor was the most significant portion of cost of goods. Overhead, at that time, was the smallest of the three cost elements, with materials less than labor but significantly more than overhead. Material costs are easy to measure as is direct labor—both are applied directly to the product and can be reported and costed. Overhead is indirect. It includes all the support costs such as the cost of the factory building and facilities, the administrative and personnel costs, inventory-related costs, maintenance, and so forth. Traditional cost accounting applies a "tax" to direct labor to absorb these indirect costs.

In the old days, the overhead tax amounted to a few percent. Today, with automation, direct labor is typically less than 10% of cost of goods, material is often 60%, and the remaining 30% is overhead. Applying overhead as a tax of direct labor requires a 300% burden. As a result, the accounting system in essence applies 40% of the cost to labor through its actual cost plus the burden. This distorts management's view and encourages a disproportionate focus on labor management and labor reduction.

Admittedly, there is the other factor: visibility. Direct labor is a lot easier to see and do something about than is overhead.

The result of this focus on labor (efficiency and the related measurement of utilization) is a tendency to keep people and machines busy even if the result is increased inventory and, more importantly, reduced flexibility. We spend millions of dollars and invest man-years of effort to implement management information systems that tell us what we should produce and when, then we destroy the benefit of these systems by driving behavior with obsolete measurements like efficiency and utilization. Make no mistake, people respond to the way they are measured. That's what measurements are for, aren't they? If the measurement drives behavior that is not tied to the system's recommendations, then the system is useless because it is unused.

By the way, there is a new approach to cost accounting called Activity-Based Costing (ABC) that was developed to address this labor-to-overhead ratio problem. In practice, properly implemented, ABC can provide a new framework for the management of support services and costs that has been used to dramatically reduce "overhead"—by as much as 80%. And these are real cost reductions, not just shuffling the names and categories they are reported under. This expansion of the basic ABC philosophy is now called Activity-Based Management.[1]

---

[1] There is a chapter on Activity-Based Costing and Management in my book *MRP+* (Industrial Press, 1993).

Who Owns Finished Goods Inventory?

A very common inventory scenario involves a cycle in which inventory increases over time until someone (usually accounting) complains that it is too high. An effort is made to reduce inventory, but shortages increase as a result. To reduce shortages, inventory is increased until someone complains, and then the cycle repeats.

This is a relatively common situation. There is a direct relationship between inventory level and customer service (an inverse relationship between inventory and shortages) that drives many companies through a repeating cycle: decrease inventory (to improve the company's "bottom line") and suffer increased shortages; in order to reduce shortages, increase inventory. See Figure 12-1. This cycle continues until and unless something changes. There was a sign hanging in the computer room of this same company discussed in the earlier scenarios that read:

> **If you continue to do things the way you've always done them, you will get the same results you've always gotten.**

The inventory situation was a case in point. In order to break the cycle, there had to be some change in how inventory and/or customer service was managed. Accounting may complain about high

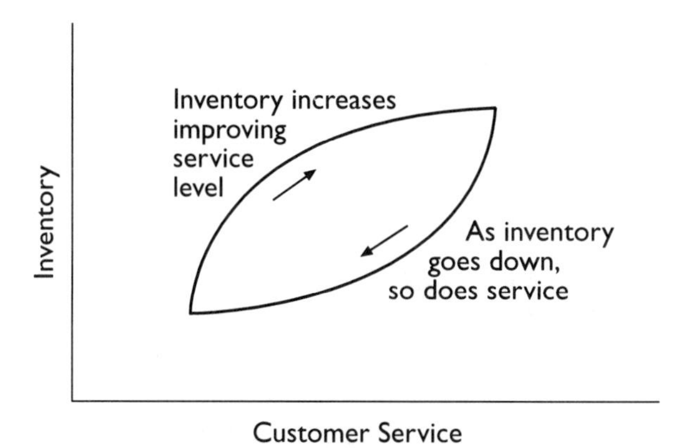

Figure 12-1. Inventory and Service Level Are Interdependent

inventories, but purchasing, planning, and sales and marketing[2] are driven toward higher inventories to maintain good service levels for their customers and hopefully increase sales (their primary measurement is sales volume because they are paid on commission).

The way it works in many companies is that the sales/marketing people provide a forecast that is used in preparing the production schedule, inventory budgets, and so on. Given the relationship between inventory level, product availability, and sales volume, it is in the sales/marketing department's best interest to be optimistic—get as much inventory as possible so it will be available if a customer orders it.

Once again, the day-to-day measurements (being able to say "yes" to customers and earning commissions on these sales) overrule any management directive there might be toward reduced inventory.

Assuming that you acquire materials and/or start production before the customer order is booked (total lead time to produce the product is greater than what is quoted to the customer), you must rely on a forecast at least to acquire materials and possibly start production.[3] If the forecast is 100% accurate, you can expect to have no finished goods inventory ... as soon as the product is made, you can ship it out. One thing that everyone knows about a forecast, however, is that it is *never* 100% accurate.

The amount of finished goods inventory is a result of the difference between the forecast and reality (sales shipments). If you forecast and build 100 units but sell only 80, inventory will increase by 20 units. If you forecast 100 and sell 120, inventory will decrease by 20 (if you had at least 120 available to ship).

Because demand (and forecast error) varies from period to period, the accepted approach to forecasting and inventory management is to provide a buffer to compensate for some of the swings that we know will occur but cannot anticipate exactly. This buffer

---

[2] I use sales and marketing or sales/marketing to represent that portion of the organization that interfaces directly with the customer. In some companies, there is a distinction between these two activities, and in others, they are indistinguishable or combined.

[3] In other words, this discussion applies in every case except a pure make-to-order situation.

takes the form of safety stock that results in some extra stock either on the shelf or in process, to compensate for variations in demand within the lead time needed to fully react.

To recap where we are, inventory allows us to ship in less than total production lead time and also helps compensate for swings in demand. The question, now, is who is responsible for this inventory?

In many cases, materials or production is considered the owner and takes the heat when customer demands cannot be met (shortages and backorders). In fact, however, you have just seen that it is the accuracy of the forecast and the amount of buffer required to compensate for the inaccuracies that really determine the inventory level. The group responsible for the forecast, therefore, should also take responsibility for the level of finished goods inventory.

Who should be responsible for the forecast? There should be only one answer here: Sales/marketing is the only group that has the knowledge of the marketplace that is required to make and manage a forecast. Unfortunately, it often refuses the assignment or is left out of the process. Or, as illustrated earlier, it doesn't understand the dynamics or is driven to unrealistic forecasts by its measurement system.

By making sales/marketing responsible for finished goods inventory, the motivation changes from complete unconcern over high inventory to a focus on having the right amount of inventory to support the business level and the accuracy of the forecast.

The "Customer Service Triangle" illustrates the relationships between inventory level, customer service, and forecast accuracy. See Figure 12-2. These three elements are tied together such that you cannot change one without changing at least one of the others. To increase customer service, either increase inventory or increase forecast accuracy. If the accuracy of the forecast can be improved, either customer service will increase (with the same inventory investment) or inventory can be lowered with no decrease in customer service. Lowering inventory will reduce customer service unless forecast accuracy can be increased enough to compensate.

Assigning responsibility for finished goods inventory to sales/marketing and creating measurements that encourage higher forecast accuracy will serve to reduce inventory and/or increase customer service.

These examples illustrates the interactions that must be addressed in an integrated system and how inappropriate measure-

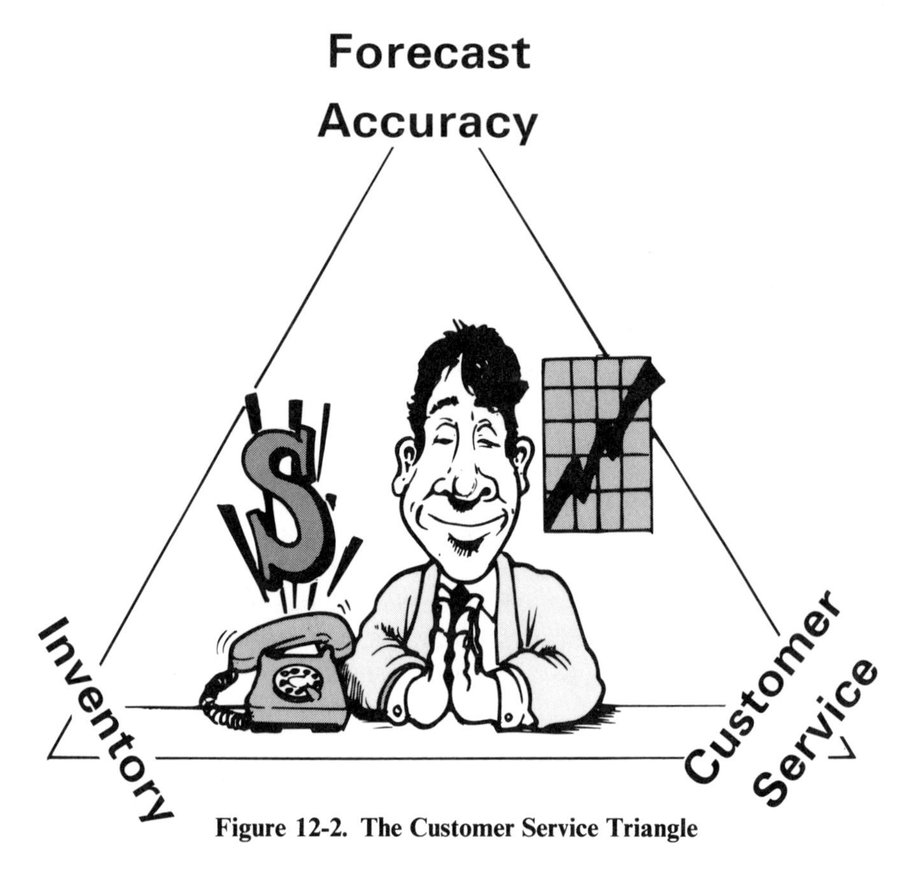

Figure 12-2. The Customer Service Triangle

ments can force people to unintentionally torpedo the expected benefits.

What Are We Measuring?

Manufacturing companies have spent billions of dollars on information systems and there have been billions of dollars of improvements, but there is a lingering feeling that the return could have been and should have been better. There are many cases, in fact, of successful implementation, ones in which all applications have been installed properly, procedures are being followed, and by all measures there has been improvement in control and performance as a result of these efforts. Yet, it seems like the change should have been more dramatic, the returns more impressive, the changes more profound.

It seems that, even if everything has been done "by the book," something was missed in the process.

Sam is the lathe department supervisor at Spin Industries, a company that has recently completed the implementation a new MRP II system. It is the twenty-fifth of June and Sam is reviewing the work list and setting the production schedule for his department for the next few days.

Sam loves the work list that he gets every day from the new system. It lists the running orders and when they should be completed. It lists the waiting orders (with priority information, so that he can decide which orders to do next on which machines). And it lists orders not yet in his department but expected to arrive there in the next four days so that he can be ready when they do arrive. Best of all, Sam likes the way management has gotten off his back.

Prior to the arrival of the new system, Sam loaded the machines as best he could based on intuition, customer due dates, and how much grief he was getting from management, sales, or customer service to get a particular job done ahead of its turn. It seemed like he could never guess right. He couldn't count the number of times that he had just completed a complicated setup and started a long run on a part when a rush job would suddenly appear and he would be forced to stop the run, tear down the machine, set it up for the rush job, tear it down again, then reset it for the job that was interrupted. What a waste, and it sure didn't help his bonus check (based on production quantities).

Although this kind of thing hadn't gone away completely, there was a lot less of it now that MRP was planning everything. Now, when some idiot came running in with a rush job, Sam would show him the work list and tell him, "If it ain't at the top of this list I ain't doin' it." He absolutely loved the look of frustration and defeat as the person turned around and headed for the production control desk. Let somebody else deal with these bozos. He still had to take care of rush jobs, but not nearly as many as before.

So, back to the task at hand, the work list for the last week of June had the usual suspects on it: some easy stuff and some jobs that would not go well, some long runs and a few short ones.

According to the priorities, the next seven jobs included a short easy one, a short tough one, a medium-long tough one, and four of the average variety. Farther down the list, there were a couple of gems—easy jobs with fairly high volumes. The priorities indicated that the easy stuff should wait until the higher priorities were finished. Sam started to scratch some notes in the margin of the work list, slotting the first seven jobs in for Tuesday, Wednesday, and Thursday. Then he stopped in midscratch and stared at the calendar. "Damn," he muttered.

Sam kept a running count of production volume for the month on his wall calendar and what he saw did not make him happy. His bonus plan was based on pounds of product produced and he was sitting at less than 70% of his target with 80% of the month behind him. The jobs he was scheduling for the next four days didn't come close to meeting his target and he was counting on that monthly bonus.

He looked back at the work list and started adding up pounds of product in different combinations. After about ten minutes, he made some new notations in the margin of the work list. The long jobs would go first, getting the poundage up to quota by late Wednesday or early Thursday, if all went well. Then he could slide the high-priority jobs in and get some of them done by the end of the week. The rest would have to wait.

On Wednesday, Sam got a call from the machining supervisor. "Where's that #3057 order that was on your schedule for Monday? I've got a machine all set up and ready to go, but today's list says you've still got it. What's the story?"

Sam made his excuses and promised to get the job over to milling the next morning. Was he upset or overcome with guilt and remorse? Not at all. This was, after all, business as usual and department supervisors were making excuses to each other every day. They were just playing the game and, this week, Sam's scorecard showed a winning hand.

What did Spin Industries' scorecard look like? Probably better than it would have been without the system in place, but not as good as it might have been if Sam and his peers were following the system-generated priorities. The prioritized work list resulted in a much smoother flow of work through the plant and reduced unnecessary setups, saving money and improving efficiency. On the other hand, the work that Sam postponed in order to satisfy his measurement system (bonus) resulted in waste and inefficiency in other departments and might well have contributed to late shipments and lost business. Certainly, customer satisfaction and Spin's reputation for on-time delivery were endangered.

Can Sam be faulted for his decision? I think not. Despite the training and instructions regarding use of work list priorities, Sam's primary measurement had to take precedence and that measurement told Sam to give poundage priority over other considerations.

Unfortunately, the preservation of obsolete measurements like this one is a major cause of failure to secure the full benefits of a new system. Most of the time it is not intentional. It is far too easy to overlook things like bonus calculations when planning a system implementation.

Mary and all of the buyer/planners at Spin Industries were just finishing a meeting with the company controller. In truth, it was more of a lecture than a meeting. The controller had gone on at great length about how inventory was much too high, the company couldn't afford to maintain that level of inventory, the buyer/planners weren't doing their part to help the company through tough times, and so on.

Mary sank down into her chair and sighed—with relief. This semiannual dressing down had been overdue and now that it was over, things could get back to normal. Oh, Mary would do her part to reduce inventory, as would all the other buyer/planners. They'd reduce some purchase quantities, hold out a little longer on some things that they would rather get in a little early, cut back on safety stock, and run things just a little closer to the edge.

All of the buyer/planners knew what would happen, however. The inventory reduction effort would be in effect for about a month, then the phone would start to ring. "We just ran out of X, get some in here right away." "The last shipment of Y had too many rejects. We need more—now." "The delivery of Z was supposed to be here yesterday, it hasn't shown up yet, and there are no spares—help!"

The semiannual inventory reduction lecture was painful and humiliating, but it was only twice a year. Shortages happened every day. A buyer/planner was responsible for making sure that needed materials were available. When there were shortages, the buyer/planner wasn't doing her job. One way to help ensure availability was by having some extra inventory (safety stock) or by padding vendor lead times (give the vendor a due date that is earlier than the real need—this compensates for late deliveries but increases inventory if the vendor delivers on the due date).

Is Mary wrong? Once again, she is only responding to her measurement system. This time, the measurement is informal. She gets grief every day for shortages, but only twice each year for excessive inventory levels. Obviously, she is strongly drawn toward the less painful alternative (higher inventory).

If she has a bonus program (formal measurement), it is most likely tied to purchase price. She is driven toward an adversarial relationship with her suppliers, trying to squeeze them for every penny of price reduction possible. The vendor has to reduce costs to be able to stay in business under severe price pressure. He, thereby, may be driven toward lower quality, and poor service. The less reliable the vendor, the more need there will be at Spin Industries for buffers like safety stock and padded lead times.

How Do You Measure Success?

Implementing a business control system such as MRP II requires investment of capitol, hard work, dedication, education, data loading, work disruption, and so on. Why do we do it? Obviously, someone must have expected sufficient resulting benefits to justify the cost.

Some benefits are obvious and easy to measure, for example, inventory reduction. Others are more subtle but can represent as much or more of the benefit of a successful implementation as the obvious things ... for example, improved customer service. Measuring and reporting these benefits are an important part of your system project. Not only is it a way to rationalize the investment (after the fact), but also as intermediate measurements as part of a long-term system improvement project.

Inventory

When looking at the inventory control area, we may want to see a reduction in inventory level as a result of our efforts. In fact, however, the implementation of an automated inventory tracking (accounting) system will not in and of itself reduce inventories. Implementing a new system, however, may be the vehicle through which we can improve procedures, practices, and accountability in the materials area, which, as a by-product, might help reduce inventory levels.

Better tracking and improved visibility can provide management with the tools needed to make better decisions. If we can now better see where things are, formalize acquisition strategy, and identify opportunities for improvements in our material management policies, then sound management decisions, indeed, can reduce inventory levels.

How can we measure success in the inventory area? The first thing I would look at is accuracy of the inventory records. Most easily identified (and improved) through an aggressive cycle counting program, the accuracy level has a direct bearing on decision-making ability and potential improvements. Another indicator of success is the quantity of obsolete inventory in stock. Good procedures and accurate information will help avoid unnecessary purchases, too-high stocking levels, and unpleasant surprises.

Does it make sense to look at inventory level at all? There's probably no way to avoid it (it's just too easy to see and too easy to relate to ... besides, the accountants will insist), but don't be misled. If you must use inventory level as an indicator, look at relative inventory measures (percentage of sales, inventory turns) and trends over time and avoid a direct link to the inventory accounting system.

MRP

O.K., raise your hand. Who installed MRP to reduce inventory? Now, did you get that hoped-for reduction? Right away?

Most MRP installations are justified, at least in part, on inventory reduction, and most experience real improvements in this area (typically, 5% to 20% or more). The *real* purpose of MRP, however, is to increase availability of parts (Master Scheduling is concerned with availability of products), and not necessarily the reduction of stock levels. In fact, most MRP implementations experience an *increase* in inventory levels during the first months of MRP use.

Chances are, before MRP is installed, parts shortages and expediting were a way of life. With MRP in effective use, those things that were being expedited (and thus never made it into "stock") will now arrive not only on time, but probably ahead of time because lead times are probably inflated. Not only that, but there will be some items in stock that shouldn't have been there at all and it will take some time to use them up or write them off.

Most new MRP implementations see a short-term increase in stock levels, which settle back to the pre-MRP level over the first six months or so.[4] Inventory reductions after that point will depend on (1) how bad things were before, (2) how aggressive (and successful) the implementation effort is, and (3) how well planning lead times are managed and how quickly lead time buffers are removed so that parts come in "just-in-time" rather than "just-in-case."

What are good measurements for MRP performance? Focus on availability, not inventory level. Measure shortages (stock-outs), and reductions in expediting activity.

Production Control

If production is continually dealing with parts shortages, don't try to judge production control system performance. The purposes of production control applications are to help manage work flow and optimize use of production resources. The disruptions caused by parts shortages will invalidate any priority or scheduling system you may be using.

Assuming parts are available, the primary goal of an integrated production control application should be to coordinate production activity to the Master Schedule. This philosophy ties day-to-day activities to customer shipments or finished goods inventory requirements (the reason you're in business).

Efficiency measurements must recognize this limitation. Additional setups, short lots, and shifting priorities caused by parts shortages will thwart any efficiency improvement efforts.

Measurement of "standard hours" earned each day/week/month or units completed in a given period will actually work against your efforts to coordinate shop activity to customer service objectives. The most important and often overlooked link between the customer service objective represented by the Master Schedule and day-to-day shop floor activity is the priority information generated by the system (MRP and the plant-floor scheduling subsystem) and how these are reflected (responded to) in actual shop activity.

Supervisors and forepersons must be given visibility of these system-generated priorities (typically through a daily work list) and (pay attention, here's the important part) be *properly motivated* to

---

[4] Of course, this interval will depend on how aggressive the implementation effort is, the amount of follow-through, and the level of discipline in applying MRP principles.

coordinate their activities to the priorities. A supervisor whose bonus is tied to units produced is not tied into the system.

Measuring the chargeable hours against capacity (utilization) or against the "standard" hours (efficiency) will not tell you if the right jobs are being completed at the right time. Setting effective measurements here is not so obvious or easy to do. The best area to focus on is on-time completion of work orders. If there are capacity problems or other extenuating circumstances, a measurement must be devised that reflects how well the plant floor is responding to the system-generated priorities. Be careful not to penalize shop personnel for things that they cannot control. Make the measurement system responsive but fair. The Japanese approach is to encourage teamwork and effort toward companywide goals by tying pay and benefits to *company* performance rather than individual performance. But the entire environment and philosophy are different there as compared to Western companies.

Purchasing

Purchasing should be judged on vendor performance in the areas of quality and on-time delivery. The price paid for goods and materials (purchase price variance or budget-related measurements) is a secondary consideration. A critical part of any just-in-time program is the development of vendor relationships to support shorter lead times, more flexible schedules, increased vendor and product reliability, and 100% acceptance levels. This cannot be done in an adversarial relationship between vendor and customer that places undue emphasis on unit cost.

Buyers and buyer/planners must be judged on their ability to develop vendors who can perform to these criteria, not strictly on cost savings.

Master Schedule

A successful Master Schedule (Master Production Schedule, or MPS) is one that accurately reflects customer demand, effectively utilizes plant resources, and does not change within product lead time. Deviations from these ideals should be the basis of your measurement system for the scheduling process itself. The ability to comply with the requirements reflects both the viability of the schedule and the underlying capabilities and disciplines throughout the operational side of the business.

Accurately reflecting customer demand is the key to controlling finished goods inventory for a ship-from-stock company, therefore, finished goods inventory levels (relative to customer service achievement) is a good measurement of forecast accuracy and Master Schedule validity. Plant or critical resource utilization reflects "reasonableness" of the MPS. A plan that fully meets customer demand but exceeds capacity will not be achieved, will build inventory, will increase lead times, and will result in poor customer service.

A stable MPS is critical to planning at the component and plant-floor levels. Here we can clearly see the conflicting goals of the MPS process: It must be flexible enough to track changing demand but stable enough not to disrupt production. When viewing lead time for a product with a multilevel bill, it is easy to see that the closer you are to product completion, the more disruptive and costly it would be to change the schedule. MRP/MPS theory designates a close-in period called the "frozen zone" during which no changes are allowed to the MPS. Beyond the frozen zone, out to Cumulative Material Lead Time (CMLT), there are other "zone(s)," sometimes called "firm" or "flexible," that represent time during which changes are less disruptive but should still be minimized if possible.

Although it would be good theory not to allow changes within the frozen zone, as a practical consideration, you cannot always be that inflexible. A good measurement of MPS stability is the number of changes within these various zones. Track the changes for a time, and then set goals for reducing the number of changes, especially in the frozen zone. Production control will thank you with higher efficiency and better on-time production performance. A back-door approach to stabilizing the MPS within CMLT is to reduce CMLT (lead time). Lead time reduction is a worthy goal unto itself, butis not usually tied directly to MRP II implementation. MRP II, however, can help identify and track the components of lead time and many MRP II–related activities will also help reduce them.

The ultimate goal, and therefore the best measurement of success, is your on-time shipment performance. If your performance in this category is high, many of the supporting processes are probably, but not necessarily,[5] working effectively. This doesn't necessarily mean you are making money or operating efficiently or that there is not room for improvement. The important thing to remember is that any changes or improvements made anywhere in the operational areas of the company must, at the very least, not adversely affect on-time shipment performance (customer service).

Measuring the Implementation

The purpose of implementing a new information system is to drive bottom-line business results. It is also necessary to measure the progress of the implementation effort itself, to be sure that it is finished on time and accomplishes what it set out to do.

The primary measurement for the project is success in meeting the schedule. The business case will point out the dollar savings to be gained when the system (and each of its components) is in place and functioning. Any delay in completing the implementation will mean some or all of those savings are lost—forever. Delays, therefore, are easily quantified in dollar terms. The impact of delays is even larger than that, however, because costs increase as the schedule stretches.

Each task in the project plan will have a due date, and task progress can be easily monitored against that date. Notice, however, that I said progress toward the date, not completion on the date. It is important to identify problems and resolve them before the due date arrives. If the task is due and is not completed, there is nothing you can do at that point to complete it on time. Identify the problem early enough and you have a chance to correct it and bring the task in on time.

It is just as easy to assign a value to early completion. Be willing to recognize such accomplishments and reward key contributors from the additional savings.

Of course, the project will have a budget, and company financial managers will be keeping a close eye on actual costs as compared to

[5] Customer service can be easily improved simply by increasing inventory, but this is not the efficient or most beneficial way to do it. Measurements like customer service (on-time shipment or fill rate) must be viewed in context; was it improved at the expense of some other important measurement?

the projected costs. They may also be watching actual savings compared to the business case projections.

Wouldn't it be nice to be able to report better-than-expected results from a project that is running under the budgeted cost? It could easily happen because most project justifications are conservative. Don't be satisfied with "as-expected" results. Reach farther and try to do more. If project targets are reached ahead of schedule, move up the schedule for subsequent tasks. If milestone measurements exceed expectations, raise the targets for the next phase. In dog racing, remember that they never let the greyhound actually catch the mechanical rabbit.

Measurements

- Measurements must change to reflect the new ways of doing business.
- Failure to make the measurements match new procedures and direction will preclude expected results.
- Measurements (formal or informal) must relate to the desired behavior (business results).
- Use ratios and trends over time.
- Make sure each measurement is in context—improvements in one area may be at the expense of another.
- People respond to how they are measured, whether it is good for the business or not.
- Use measurements to identify the problem to be solved not the individual to be blamed.
- Measure objectively. Don't measure first and set the target after. Watch out for perfect scores, they may indicate that this is happening.

13. Can't I Just Buy It?

In late 1989, a very sad tale was widely reported in the trade press about a law suit filed by a company called Diversified Graphics, Ltd., against a large, well-known consulting firm. It seems that the client asked the consultant to find them a "turn-key" system that would "be fully operational without need of extensive employee training." The end result was that the client was unhappy with the system, so it sued the consultant, and won. In response to one of the magazine articles describing the case, I wrote the following letter to the editor.

To the Editor:

The situation described ... where an unfortunate company was disappointed with its computer system and successfully sued the consulting firm was indeed a sad tale. In this kind of situation, no one wins except the lawyers. While the company was successful in their suit, they were probably not able to collect adequate compensation to cover all of the direct and indirect losses that resulted from their unsatisfactory computer implementation.

I realize that the point of the article was the consulting firm's liability but I feel that there is a more basic issue here. The fault of the consultant was not that he didn't deliver but that he promised the impossible. In information processing, there is simply no such thing as a "turn-key" system.

The client claimed (successfully) that the consultant promised to locate "a 'turn-key' system that would be fully operational without need of extensive employee training." This statement reflects a total misrepresentation of what a computer system is and does. Information systems are tools to be used in the day-to-day handling of business information. They are not and cannot be fully automatic. To effectively use any tool, training is required. To assert that this is not the case is patently incorrect.

From the information in this article, which I realize contained only a brief summary of the case, I feel that the suit should have been for fraud rather than professional liability. The customer was sold unrealistic expectations. The consultant couldn't possibly succeed. To accept the consulting assignment under these circumstances represents extremely poor judgment by the consulting firm.

The court's decision represents a precedent that we will all pay for in the coming years and I feel that this is very unfortunate not only for the consulting community but for users and potential users alike. Instead of focusing on "doing the right thing" and getting the job done, both sides will waste untold effort and resources protecting themselves from liability. The cost of computing has just risen significantly.

While I wholeheartedly support any effort toward improving and maintaining high ethical standards in our industry, I hate to see the focus changed from the development of better services and closer relationships with our customers to one of mutual distrust and confrontation that will inevitably result from the potential for litigation.

The article ended with some advice for consultants that included: "use care in taking on a project." I couldn't agree more. This case is a good example of what can happen when the client's expectations are not properly set at the beginning. People who are not involved in the information processing industry can easily be misled about the capabilities of computer systems. We in the industry have our own vocabulary and our interpretation of certain terms can be very different from an "outsider's" interpretation.

And by the way, let's delete the term "turn-key" from the language. It is a meaningless term to the systems professional and far too meaningful to the uninformed.

I hate the term "turn-key." This word or term doesn't appear in the dictionary on the book shelf above my desk, but most people have an idea about what it means; when you buy a turn-key system, the vendor will deliver or install it and hand it over to you, ready to go ... just "turn the key" and you're off and running.

The definition of complete installation, however, might be quite different, depending on what the product is and also on which side of the transaction you happen to find yourself. Let's start with a simple illustration and see where it takes us.

Imagine you just bought a new bottled-water dispenser. The delivery person wheels it into place, plugs in the cord, and installs the first jug of water on the top. I believe most people would be able to get a glass of water from the dispenser without further training. If the dispenser is a deluxe model with both hot and cold spigots, it might take a little experimentation to figure out that the blue one is cold and the red one hot, but other than that, I think you're O.K. Turn-key installation for a water dispenser simply means installed and operating. Now let's try something a little more complex.

We just got a new copier in our office. The installer set up the machine and its accessories (sorter, large capacity feed bin, and so on) and spent about a half hour showing us how to use it. In this case, the installation was obviously complete when the machine was set up and tested (operating), but if the installer had left at that point, we could not have put the machine into full productive use without spending considerable time studying the manual (assuming, of course, that the manual was complete and understandable ... not always a good assumption) and without some experimentation. After the *installation* was complete, we had to go through an *implementation* process (a brief training session, in this case) to make the new machine useful. Even at that, we had a bit of a learning curve on some of the less obvious features of the machine such as clearing paper jams and refilling the toner supply.

With a complex system, there is an important distinction between installation and implementation that is ignored by the term turn-key. Implementation invariably involves active participation by the future users of the system. It is one thing to have a system (computer, appliance, production machine, whatever) installed and ready to work ... but often quite another to have it and the supporting liveware (people) in a position to fully exploit its potential benefits.

Survey question: Does your VCR at home continuously flash 12:00, 12:00, 12:00?

The VCR may be properly hooked up to your TV and cable or antenna (fully installed). You may be able to record a program while you are watching it or play prerecorded tapes. But, can you program it to tape three different shows on different channels on different days while you are away on vacation? Even if you bought the VCR with "installation" included, would you normally expect to be able to fully use it based solely on intuition (without training)? If the store promised a turn-key installation, what level of utility would you expect when it handed you the remote control and drove away? How much of your own time would you have expected to invest in learning how to use the machine?

This cited court case has wide-ranging implications in the area of consultant liability, but the point that I would like to make is this: The consulting firm didn't deliver because it couldn't. The client expected the impossible. There is no such thing as a turn-key computer system that requires little or no training. If something as simple as a home video recorder or small office copier requires training and experience to use effectively, how can you expect to *intuitively* know how to operate a system that is intended to track an entire company's information?

I remember the first time I was introduced to the System/34, a now obsolete IBM minicomputer that was very popular in the early 1970s particularly because of its outstanding ease of use. I was a consultant on a new installation (I had assisted in the system selection process), and on the first day that the system was installed, the IBM systems engineer sat down at the console and ran a quick utility procedure to fix a minor problem with a file. I believe I am a reasonably intelligent person, and I had considerable systems experience at that time, but those screens flashing by made my head spin. Despite the fact that the systems engineer was narrating the procedure, I was completely unable to follow it because it was entirely new to me. After having worked with System/34s for a number of years after that day, operation of that system and use of its utility functions became second nature to me, but I can still remember how mysterious and intimidating they were when I first saw them.

For me, an IBM AS/400 system (a several-generations-removed successor to the System/34) could be considered installed when it is physically in place, plugged in, and the operating system is loaded and ready to go. I have enough experience and familiarity with the machine to be able to operate it without further training. Not so for a new user. No matter how "user-friendly" the machine might be, it is absurd to assume that someone can simply "walk up and use" it, as some software vendors assert. I also don't presume to know how to run any particular application program that I have never seen before just because it happens to run on a box that I am familiar with. And, even with software that I know well, I don't necessarily know how it will apply in a particular company's situation until I know something about the company and how it operates.

So, the definition of turn-key will be different, depending on the experience of the "customer," as well as the extent of the installation activities provided.

A company effectively using one MRP package, for example, would need very little training to move from one particular release (version) of that package to the next release of the same system because the functions will undoubtedly be basically the same (with improvements) and the real implementation tasks—putting the information to use—were accomplished on the initial effort. A company new to MRP II, or a company moving from one brand of MRP system to another, would need sufficient training to understand the new system's functions and must also weave the new system's functionality into the business and processes of the company.

The users must be trained, procedures adapted to take advantage of the system's capabilities, discipline must be imposed to assure data accuracy, and on and on. Notice that all of these considerations are things that the company must take on themselves ... they are not things that can be delivered by a vendor or a consultant.

In the world of information systems, the measure of success is not the flashing lights and spinning tape reels in the computer room. Successful implementation is the effective day-to-day use of the system's facilities to enhance the operation of the business through more effective control and more informed decision making. Any vendor or consultant that guarantees results or promises to "do it all" for you is either hopelessly naive or is being less than forthright with you. It cannot be done by outsiders. Only *you*, the user of the system, can implement it. The vendor can only install, convert, assist, and advise.

Of course, experienced advisors can be a big help. Just because the responsibility is yours doesn't mean you have to reinvent the wheel. Those who have been through similar implementation projects before can be of tremendous help in pointing out the best way to do something, helping you avoid missteps, and making the most of your investment. Use these resources as much as you can, but don't be tempted to off-load portions of the project on your outside resources.

Using Consultants

The use of a consultant can be a key element of a successful systems project. It can also be a complete waste of money or actually an impediment to effective implementation. A lot depends on the consultant's skills, of course, but even more important is how this resource is used or misused by the client company.

> Consultants must be a good thing . . . else why would there be so many of them and why would we pay them so much money!
>
> *Anonymous*

There are several ways in which a consultant can be used in a system implementation project: up front to assist in planning the project; during initial implementation to provide expertise and guidance and to deliver some of the education and training; and as a continuing resource to assess effectiveness, advise on improvement ideas or extension of system use, and for problem solving.

Up-Front Planning

An obvious use of a consultant is to benefit from his or her experience of having "been there." The consultant you choose for the planning phase must be one who has experienced at least one and hopefully many previous implementation projects similar to the one you are about to undertake. It is hoped that in previous projects, their experience has taught them how to "make it happen" and can also help you avoid some of the common errors and false starts. The consultant's ability to bring that experience to your situation and relate it to your project in a way that will help you effectively implement it is the major benefit she can provide.

The biggest risk in the implementation planning process is in letting the consultant plan your project for you. Sure, it would be much easier for the consultant to lay out a plan than for you to do it yourself and may, in fact, have a plan already prepared (borrowed from previous efforts) that you can just "plug in" and use as is or with minor changes. Don't do it!

The essence of effective system use is in the users' sense of "ownership" of the project and the system. In order for the hard-working employees of your company to become completely comfortable with the new system, and to devote the extra effort necessary to get it installed, loaded, and woven into their daily activities, they *must* be involved and committed from the beginning of the project.

You cannot expect to present a completed project plan to your employees, ready for their action, and expect wholehearted commitment. The project team and the people who will be doing the implementation tasks must be involved in the development of the plan and schedule so that they understand what they are signing up for and feel that it is *their* plan, not something that was forced upon them.

A consultant can provide guidance in the planning process, can advise on sequence and dependencies, and can help develop manpower and timing estimates, as well as critique the plan that you develop, but the work must be yours.

Implementation Stage

Perhaps the most common use of a consultant is for guidance and education during the actual implementation stages of the project. An experienced "expert" can point out common pitfalls, recommend successful approaches, confirm the appropriateness of decisions that you make along the way, help make you aware of the future impact of early decisions (especially related to loading data fields, application tailoring, and procedure development), relate "creative" uses of the system's facilities that were developed or witnessed at other companies, and generally help in keeping your project on target.

Here, also, the greatest danger is in having the consultant do too much. The consultant must not take responsibility for carrying out any specific tasks within the project, and should not be allowed to make decisions for you. In other words, the consultant cannot be an active participant in the project ... just an advisor. Responsibility and authority cannot be delegated to anyone who is outside of the company (not a future user of the system).

The project that relies on an outside resource for task accomplishment or decisions has a built-in scapegoat for any problems or difficulties. When the project falters or gets into trouble, it is far too easy to lay blame on the outsider, thus protecting the staff from responsibility, whereas the proper delegation of responsibility is precisely what makes the project work.

During implementation, the consultant can be a good source of education and training. Their past experiences, coupled with a knowledge of your needs and objectives, make the consultant a good candidate for teaching duties. Not all consultants are good instructors, however. Some are not comfortable speaking in front of a group, others are not well organized or lack teaching skills, and many do not have prepared class materials. An experienced consultant is best qualified to present concepts and generic system implementation and use information. If the consultant has extensive experience with the particular software that you are installing, you may be able to acquire specific implementation and use topics. You may, however, have to rely on the software supplier to provide this detailed education and training. The supplier is best positioned to have prepared training materials and experienced instructors for the specifics of its package.

Expect to take responsibility for some of the training yourself. One of the best ways to reinforce new skills is to teach them to others. In preparing for your in-house training program, your instructors will confirm their own understanding of the material and they will also have to organize it for presentation, which helps confirm it in their own minds. Most software suppliers will offer teach-the-teacher training for your in-house instructors. Having in-house training resources also helps maintain the high skill level that is established during the implementation process. As people move within the company and new employees are brought in, in-house training resources can be exploited to maintain a high level of skills and understanding.

It is also recommended that you do not rely on only one source for all of your information. If a single consultant is your only teacher, you will have only one opinion and one view of the world. Even if the opinion turns out to be absolutely the best for you, it helps to have some exposure to a different experience base and different personalities. At the very least, send a few people to scheduled classes from established vendors (your software supplier, any available third-party education resource, professional societies such as the American Production Control and Inventory Society, professional training resources for your industry).

Continuing Support

After the initial implementation effort is complete, many companies will maintain a long-term relationship with a consultant for a number of reasons:

- The consultant can periodically "audit" (evaluate) the system and procedures to see how effectively they are being applied. Based on the audit, improvement activities can be undertaken.
- The consultant can assist in the planning and implementation of additional functions (new applications, features not being used) and extensions of the system.
- The consultant can help you keep abreast of technological developments (hardware advancements, software improvements, new or expanded management theories and approaches) and evaluate the potential benefits to the company.
- The consultant can be used for problem solving and continuing education.

How Much, How Often?

Early in my career as a consultant, I had the opportunity to work with a medium-sized company in some early preparations for the implementation of its new MRP system. At the completion of this first few days of activity, the company asked for a proposal for how I could support the implementation itself. Being fairly new to the consulting business, I wrestled with this proposal and finally came up with a program consisting of a number of days of education and training for the project team and users plus a sprinkling of consulting days scattered through the next year.

When I presented this proposal, the company's response was, "You're really cutting us loose—leaving us on our own. We want you here a lot more than this." I told them how strongly I believed that they must develop self-sufficiency and emphasized the education portion of what I had proposed. They weren't satisfied with this answer.

Because I was unwilling to give them what they wanted, they found another consultant who would. The consultant was on site practically full time for the first year and half time for a long period beyond that.

I still have occasional contact with this company. It has had a measure of success with its system (and the two major upgrades since the initial installation) but not as much success as it initially hoped for or, I believe, it should have had. Although the consultant did help get the system implemented, the users were less involved, felt less responsibility, and gave less of themselves to the project than was needed to be truly successful. A dozen years later, the users are still not using the system to really manage their business. Inventory control is mediocre at best. They don't yet use MRP to plan production and purchasing schedules. They have not really addressed shop-floor control.

A common approach to consulting, typical of the larger firms and the consulting arms of the large accounting firms, is to put one or more people on site (full time) for an extended period of time. A variation of this technique is an extensive schedule of visits such as two or three days per week for up to a year or more. I have always been very wary of this kind of arrangement. I believe this can be detrimental to the project because the consultant becomes too much of a member of the team. With constant availability, the team members always can go to the consultant for advice or verification, can defer decisions easily (until the consultant is here, if not full time), and avoid taking full responsibility. If the consultant is there two days per week, very little really gets accomplished on the other three days.

It is vital that the team (the company, its management, the employees) is self-reliant and takes full ownership of the project. Although sound advice based on experience is essential to success, overreliance on outside resources is not only expensive, but is actually harmful to the project.

In the early stages: during planning, initial education, and cutover, and at key points in the implementation such as "going live" with a new module, a consultant might be a regular fixture on the scene. Other than at these critical junctures, however, it should not be normally necessary to have the consultant on site for more than a day or two at a time and no more often than once or twice per month. I have several clients, in fact, (not implementing any new functions) who schedule one day several times per year similar to the way you would have an oil change or tune-up for your car. Minor questions or problems are handled by phone, and new ideas or such things as proposed procedural changes are discussed at the periodic "checkup" visits.

The kind of consultant I'm talking about here is a management consultant, not a programmer/analyst or technical advisor. Technical people obviously would take responsibility for specific tasks and would likely work on those assignments continuously until completed.

Choosing a Consultant

The most important characteristic for a consultant helping you implement SYSTEM X is a wealth of SYSTEM X experience. The more different installations that the consultant has supported, the more situations and techniques she will have observed and she can bring the expertise gained thereby to benefit in the effective completion of your project.

Look for a consultant that has knowledge and experience in all SYSTEM X application and related areas. You will make decisions early in the implementation that will have a big impact on functions that you might install much later. Be sure the consultant knows the entire product, not just the basics.

The consultant should also be familiar with the issues and considerations that are not strictly SYSTEM X–related, specifically, such things as the theories and practices of manufacturing management in general, financial management, cost accounting, personnel issues, marketing considerations, and so on. Look for professional recognition such as APICS certification as evidence of competence outside of SYSTEM X.

If your company is part of a specific industry segment, such as process, custom manufacturing, or government contracting, look for a consultant with experience and expertise in your area. You don't want to have to explain your industry from square one, although you will undoubtedly have to familiarize the consultant with the specifics of your company organization and procedures.

Finally, find a consultant that is a good problem solver, is a clear thinker, is pragmatic (you need solutions, not theories), and has good communications skills. Admittedly, these characteristics are hard to assess from a résumé or an interview. The best approach is to talk to some references. If the consultant has happy customers (in a similar industry or with similar needs) who are willing to provide an encouraging recommendation based on long term experience and confirm the quality of the advice given by the consultant, you have found a winner.

Consultant or Contractor?

So far I have addressed the need for a consultant—the expert advisor who helps plan and manage the implementation process. There are other outside experts or resources who can be engaged during a system implementation that I classify as contractors; they provide services other than implementation advice and assistance. The most common of such resources is the contract programmer/analyst.

The rules are a little different for the contractor. Specifically, the contractor is *expected* to take responsibility for the completion of a defined task, whereas the consultant should not. Contractors also are usually assigned to a project for a period of time, for the specific purpose of delivering a defined service.

Although it makes perfect sense to rent expertise that you don't have available and only need for a short time, a company employee and member of the implementation effort must take responsibility for managing the outsider resource. There will likely be a contract in place that specifies what the contractor will do and when, and someone on the company payroll will be the technical point of contact, monitor progress, act as liaison, and verify completion of the task.

There also might be contract employees or temporaries involved during implementation. Be careful in assigning task responsibilities to people who will not be the ultimate users of the system. Specific jobs can be contracted, but responsibilities cannot.

The Rule of the Tool

I recently provided some consulting services to a (Fortune-500 sized) company that was in the throes of a very large and extensive business reengineering project. Although the project covered the entire range of the company's business processes, redesign/replacement of the information system was a major part of the effort.

The company had contracted with a major consulting firm to lead the reengineering effort. The consultants did a great job of organizing the effort, leading the various subteams through problem identification and process redesign, and documenting the as-is and to-be models for the company.

After selection of the computer system and packaged software that would support this new vision, the company chose to get the majority of the implementation services and assistance from the software supplier rather than the consulting firm. The primary reason was that the software supplier could provide more focused training and assistance, based on extensive experience implementing that specific hardware/software combination. The consulting firm claimed some experience with that solution, but the majority of the consultants assigned to the project had none.

The consulting firm, at that stage, had more knowledge of the company, its organization, needs, capabilities, and politics, but lacked specific experience with the chosen solution. The fact is that the consultant would have been happier with a solution that she was experienced with, but the consultants' preferences did not prevail in the selection process.

One of the most valuable characteristics of a consultant is objectivity. Presumably, the consultant is paid to provide an unbiased view and recommendations. The client relies on this objectivity when it accepts the consultant's advice.

I was recently involved in a system evaluation for a company with two plants—one on the east coast of the United States with system A installed and a west coast operation that was using system B. The east coast plant was generally satisfied with system A, but needed some additional investment (upgrades) to extend its use. System B, on the other hand, was generally considered to be inadequate and not upgradable, and therefore had to be replaced. Corporate management wanted the same system to be used in both plants.

As I began to work with this company, the management team gained confidence in me and asked many questions about system implementation and how I could assist after the system selection process was completed. I was practically guaranteed the consulting job to help with the implementation.

It struck me that self-interest could easily cloud my judgment in the selection process. If the company chose system A, there would be a single implementation project, at the west coast plant. If an entirely different system was selected, there would be two *implementation projects, possibly doubling the opportunity for consulting services.*

Consultants are constantly faced with such ethical questions, many times not as obvious as this one. He must constantly be aware of potential conflicts between his own best interests and those of the client.

This is one reason why I have always felt uncomfortable with the idea of accounting firms, programming shops, and software vendors offering consulting services. No matter how ethical a company or an individual might be, there must be tremendous pressure to recommend the products and services of your own company, at least in part because those are the things that you know and you must believe they are good (you shouldn't be working for that company if you don't believe in its products).

There is also the "Rule of the Tool." See Figure 13-1. If you are a programmer and you encounter a problem or challenge, it is likely that the best solution in your mind will be a programming solution. If your background is consulting services, that will be the default solution. If you're an industrial engineer, procedural changes will appeal to you.

Bias can be unintentional and not necessarily self-serving. If I am not familiar with software product X, it might be more difficult for me to recommend its use over package Y that I know very well ... even if it makes no difference in future business or income for me. Even if I could anticipate more business by choosing package Y, I might still tend toward X due to its familiarity.

Can a consultant ever be truly objective, then? Probably not. But that doesn't mean that consultants cannot provide valuable advice and the benefit of their experience with others who have faced the same challenges that you now face. By all means, exploit this resource, but understand the consultant's position. A consultant is an advisor, not a decision maker. A consultant should not prepare the project plan nor take responsibility for the completion of any specific project tasks.

The RULE OF the TOOL:
If you have a hammer in your
hand, the whole world looks
like a nail.

Figure 13-1.

Turn-Key

- Information systems are tools to be used during the performance of regular duties. The users must be trained to use these tools.
- No software or system supplier can deliver a complete system that will function on its own.
- The ease of use of a system is dependent on the background and training of the user as much as on the system's design.

Consultants

- The consultant's greatest asset is her experience. She can help you learn from the successes and failures of other implementations.
- A consultant should never take responsibility for accomplishment of any project tasks or deliverables.
- Avoid overdependence on outside resources. Build self-reliance as part of your ownership of your system.
- Contractors are not consultants and therefore can be held responsible for deliverables.
- Be aware of a consultant's affiliations and loyalties. Don't assume objectivity. A consultant with vendor affiliations can still be a valuable advisor, but keep the affiliations in mind when evaluating her advice.
- Be aware of your consultant's background. "If you have a hammer in your hand, the whole world looks like a nail."

14. Final Words

The success of a system implementation is far more dependent on the implementation than on the system. As I mentioned in the beginning, I have seen companies fail with the best of systems and others succeed with the worst.

One company I worked with recently is a striking case in point. The operating division I worked with was newly formed as a result of a reorganization. There were two plants: one in the Northeast and one in the Mid-Atlantic area. The products of the two plants were somewhat different, but were sold within the same industry segment. The two plants were previously associated with different divisions within the same corporation, which had been organized by product line. Now they were thrust together as part of the North American division in the new corporate geographic alignment.

The Mid-Atlantic plant was using a well-known MRP system that had been in place for about eight years. Both the hardware and the software package were typical for that vintage, offering a complete suite of applications and functions. The plant was only using the basic applications, however, primarily inventory management, customer order handling, and the financial applications. It expressed an interest in MRP and plant-floor control, but had not pursued those interests.

The Northeast plant had implemented a new system over the last three years. It was fully utilizing the MRP and production control functions (as well as purchasing, inventory, customer order handling, financials, etc.) and would have earned very high grades in any objective evaluation of its success in fully using the system's functions.

The interesting thing about these two plants is that, despite the fact that the Mid-Atlantic plant's system was five years older, it was at least one generation more advanced than the one recently installed in the Northeast.

It's hard to know what motivated the Northeast plant to select and implement what amounts to a technological dinosaur, and there's no profit in trying to assign blame at this point. The lesson here is exactly as stated earlier: The Northeast plant succeeded despite a very limited (and decidedly nonuser-friendly) system, whereas the Mid-Atlantic plant failed to achieve anywhere near the same results, despite the fact that its system, in and of itself, was superior to what the other plant had to work with.

Hardening of the Attitudes

I recently taught a class in which there were two people, sitting side by side, who clearly illustrated what a difference attitude can make.

Marjorie, the purchasing manager at the company, was relatively new to the company and, though quite experienced in purchasing, was also new to the system that was in use at this company. Beth, on the other hand, was a long-time employee who supervised the Accounts Payable activities. Marjorie was an active participant in the two days of class, asking many questions and striving very hard to understand the package and discover its strengths and weaknesses.

This software package, widely used in thousands of companies throughout the world, was typical of any general-purpose package, that is, it performed most functions reasonably well but might fall short of perfection when viewed against the needs of any particular company. No package is a perfect fit, after all. In any case, Marjorie knew about many of the system's limitations from her experience with the package during her six months with the company but was eager to have her observations confirmed by the "expert." She was also hoping to be proved wrong. If it turned out that her difficulties were the result of misconceptions or simply lack of training, she would be extremely pleased. Although it is embarrassing to be wrong, Marjorie realized that what she learned at the class would allow her to perform better, use the system more fully, and execute her assigned duties more effec-

tively. "So what" if she didn't figure it out on her own? She was happy to be getting the information.

In addition, Marjorie welcomed knowledge about system limitations. Where the system fell short, she wanted to explore how she and the company could compensate. Was there a work-around? What was the impact of this limitation? How did other companies deal with it?

Beth had received some minimal indoctrination to the system several years earlier when it was first installed. The "so-called experts" (her words) told her to do this a certain way and do that a certain way and she didn't want to hear about any other ideas. She wasn't interested in understanding what the system did or why it was designed the way it was. She wasn't interested in discussing options or hearing how other companies had dealt with peculiarities of the package.

She was having problems with the invoice entry process. In this package, invoices were electronically matched to purchase order information and the system would produce warning messages alerting the entry person when quantities didn't match, when prices differed, and so on. The procedure required that certain things be entered in a certain way and, in reality, the process was a bit on the finicky side. Responding to the messages was pretty straightforward, but not especially user-friendly. In addition, local procedures added considerable inconvenience to the process.

The company was a division of a much larger multinational corporation and the divisions were forced to use the corporate Accounts Payable system. This required a "bridge" program that took the payables information from the local purchasing system and passed it to the corporate payables program. The bridge, though perfectly functional, was written with a limited understanding of the detailed function of the purchasing system. As a result, the invoice entry and clearing process was more cumbersome than it might have been.

Within a short time in the class, we had explored the software's basic invoice entry process and what the established procedures required. There were, in fact, several rather simple things that could be done to make things a lot easier. In addition, we were able to determine that the bridge program could be easily changed to greatly enhance and simplify the entire process. At the end of the session, I was feeling quite good about what had been discussed. Beth didn't say much at that point and, I admit, I was not closely watching her body language. Therefore, it was a shock to me when, as Beth was leaving the room, she said, almost inaudibly, "I don't know a whole lot more than I did before."

The others who were still in the room noticed my surprise, and immediately apologized for Beth, telling me that it was just the way she was. She didn't like the system and apparently didn't want to like the system. She, in effect, rejected the suggested improvements because they would alter her position as long-suffering victim. I guess there's just no helping someone whose mind is sealed closed like a bank vault at midnight. Unfortunately, such people become immovable obstacles in the way of progress toward effective system implementation. Beth staked out her position as victim and resolutely remained in the middle of the path as the rest of the world tried to keep moving ahead around, over, under, or through her.

Marjorie, on the other hand, had some problems that were, quite honestly, a lot more serious than Beth's. It would have been very easy for Marjorie to view the package as completely unacceptable and become an obstacle, as Beth had done. Instead, she actively sought ways to use what was good about the system and explore ways to work around what was not so good.

As you begin to plan your system implementation, keep in mind that it is the "people" issues that will be the prime determinants of success or failure. Much thought goes into the selection of hardware and software because they are visible, easily priced out, and you can see and feel what you get for your money. The investment in the human side—implementation effort, training, work disruption, procedure development, and the rest—is a lot harder to get a handle on.

Figure 14-1 appeared in a report I wrote for the Yankee Group, a Boston-based market analysis firm, on trends in MRP II. The numbers behind the chart are my own estimate, based on years of experience and not any one particular implementation.

If anything, the hardware/software numbers may be too large relative to the other costs in this figure. In addition to the size of the "soft" costs, notice that there is a fairly large slice dedicated to "maintenance." This estimate is supposed to be for a typical new system implementation for the life-cycle cost of the system, that is to say, the total cost of ownership over some specified period of time—perhaps five years. The maintenance costs in this chart may be considerably understated, depending on how maintenance is defined (does it include continuing education or retraining, support costs

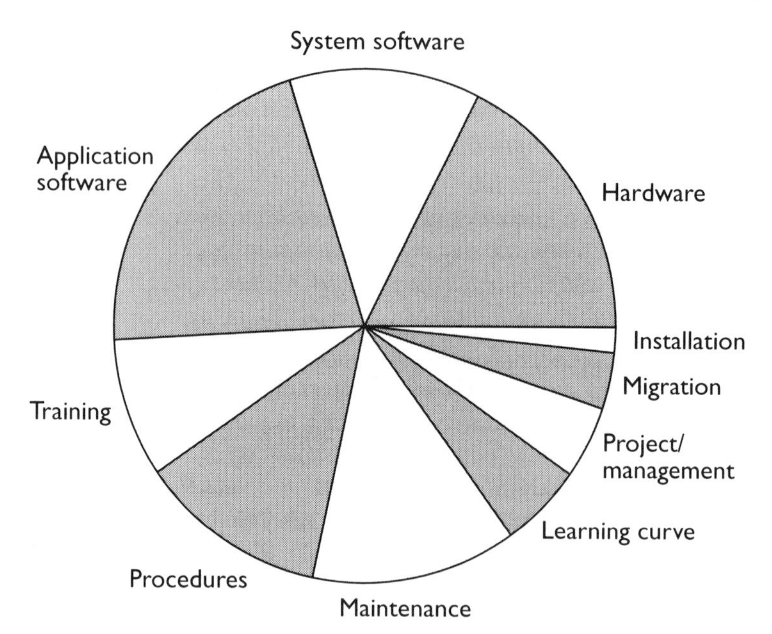

Figure 14-1. Relative Cost of System Implementation Elements

that are buried in operational budgets, and so on?—probably not) and the length of the projected "life" of the system.

Key Elements

To reiterate, the key elements of success in a system implementation are as follows:

- Executive Commitment: not just involvement, but full, continuing, and active commitment to the successful completion of the project.
- Organization: including the organization that implements the system and the changes to the company organization that are necessary in order to reap the benefits.
- Education and Training: there is never enough. My best advice is: Whatever you budget for education and training, double it. It will be the best investment you will make.

The Oliver Wight company conducted a survey of companies that had implemented MRP II systems (about 1,100 responses) and found a number of interesting facts including the following:

- The total amount spent for a system and its implementation was nearly the same for very successful companies as for unsuccessful ones.
- The successful implementations, however, spend several times as much on education and training.
- The one thing that just about most respondents said they would do differently was "more education"—even the most successful ones.

We're Already Busy

In today's business world, nobody is standing around looking for something to do. Every company I've worked with in the last fifteen years is running "lean and mean" with no extra resources to assign to the system implementation project. And yet, implementations are completed, somehow the resources are made available, and the other regular jobs are done at the same time. How is this miracle accomplished? If it is important enough, it will be done.

The trick is to make it (the implementation) important enough to each individual so as to enlist his or her extraordinary effort. But extraordinary effort is only available for a limited time. Endless projects never end—if the project duration is too long (to get the extra effort), it will not be successful. In fact, the longer the project takes, the less like it is to succeed and the more it will cost.

Some of the following can help:

- Make sure the duration of the project is reasonable and have the "volunteers" to sign up for the duration. Then make sure you complete it on time.
- Have the chief executive deliver a personal plea—describing the importance of the project to all concerned and pledging his or her full support—and deliver that support for the duration of the project.
- It's O.K. to bring in temporary help. While new tasks are being added and old ones have yet to go away, use the "temps" to keep the old processes going. Let the regular employees concentrate on learning and using the new processes.
- Postpone or suspend lower-priority tasks.
- Be understanding. There is a lot of extra effort required and a lot of encouragement is needed.

Appendix A
A Brief Overview of MRP and MRP II[1]

"The first thing a manager should know about MRP is that there are two of them. Usually referred to as MRP and MRP II (or Little MRP and Big MRP), they stand for Material Requirements Planning and Manufacturing Resource Planning, respectively. MRP is a well-defined process (set of calculations) that is used to develop plans for the acquisition of the materials needed to meet the needs of production. MRP is often embodied in a software module or program as part of a larger manufacturing management information control system.

"MRP II is an information control philosophy, often translated into computer software products, which contains, among other capabilities, an MRP calculation function. MRP II includes the integration of production operations (inventory, production control, purchasing), production planning (MRP, capacity planning, master scheduling), customer service (order entry, sales analysis, forecasting), and financial applications (general ledger, accounts payable and receivable, payroll) into a single information control system that shares data among the various applications for their mutual benefit.

Material Requirements Planning

"MRP refers to a specific technique for generating recommended material acquisition activities (purchase orders and production

---

[1] Appendix A originally appeared as "What Every Manager Needs to Know About MRP II" by David Turbide in the January 1990 issue of *Manufacturing Systems*. Copyright 1990 by Chilton Publishing Company.

activities) to meet future demands. These demands are specified in a Master Production Schedule (MPS), which is a statement of planned manufacturing activities for the items you sell (products, end items, service parts). Starting from the Master Schedule, MRP goes through a four-step process to determine needs at lower levels (of the bill-of-material). See Figure A-1. The four steps follow:

1. Determine the components needed by the MPS planned orders. The start date of the planned MPS production order is the need date for the components. Component need quantity is determined from the order quantity and the single-level bill-of-material

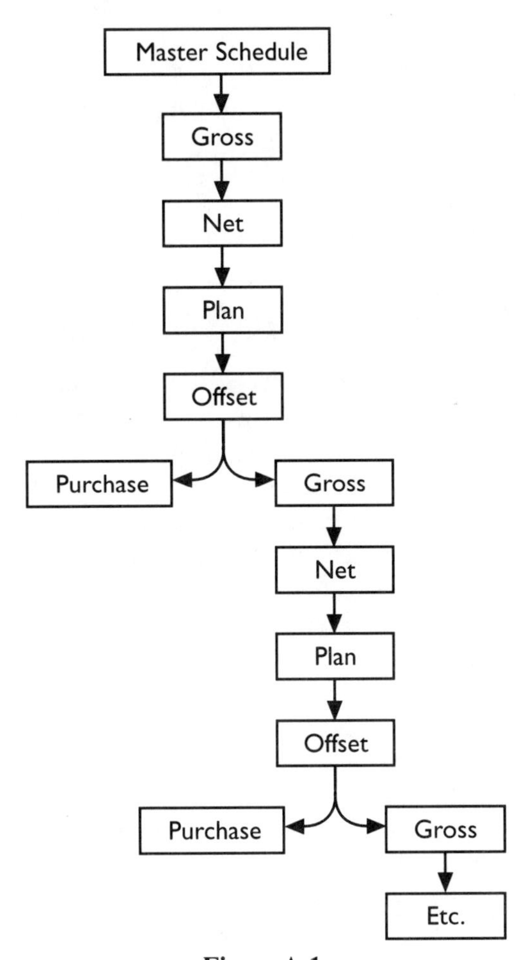

Figure A-1.

information, which must be stored in the computer. The resulting component needs are called Gross Requirements.

2. Once Gross Requirements are determined, the system checks expected inventory availability for each component on the need date. Available inventory quantity is equal to today's on-hand balance less any expected usage (allocations) plus any expected receipts (on-order quantities) between now and the date of need. Any expected shortages thus identified are Net Requirements.

"Also in this step, the system will attempt to align any open orders to the needs. If an item is expected to have a shortage and there is an existing order for the item with a due date later than the need, the system will recommend that the order due date be moved up (expedited) to match the need date. In like manner, for open orders with due dates too early (item not needed until later), MRP will recommend deferral. The netting step assumes that you will expedite and/or defer as specified and will proceed accordingly.

3. The third step is to plan acquisition orders according to any preset lot sizing rules that may have been imposed for the item. For example, it may not be desirable to make something in less than a certain quantity, or, because of packaging considerations, something is always purchased in dozens or hundreds. Most MRP systems provide a number of methods for lot sizing, which can vary from item to item.

4. Finally, once order sizes are set, the start date for the planned order is determined by subtracting the lead time (production lead time or purchase lead time, as appropriate, which also must be stored in the computer) from the need date (due date).

"If the item thus planned is a purchased item, the process ends here with planned purchase orders. For manufactured items, the process is repeated for *its* components, and *their* components level by level until a complete material acquisition plan is developed.

"The end result of the MRP process is a series of planned orders that are timed to bring in the additional needed materials just in time to be available for higher-level production activity.

"This process, in theory, could result in perfectly timed production and purchase orders, for 100% availability (no shortages) and zero inventories. In order to achieve perfect MRP, however, you would have to have 100% accurate bills, 100% accurate inventory records, never have any unanticipated scrap or other losses, and never miss a due date.

"In the real world, therefore, MRP will not result in zero inventory and no shortages. Because we know that nothing's perfect, we buffer ourselves in several ways to make up for our less-than-perfect planning information. The better we do with the prerequisites (bills and inventory accuracy, meeting due dates), the more we can reduce inventories and improve availability (avoid shortages) at the same time. Reduced shortages is a much better measure of success with MRP than is inventory reduction.

"Inventory is a buffer against difficulty in planning, variations in demand, quality or production problems, failure to meet due dates, and changing schedules.

"Lot sizing policies, safety (stock) requirements, shrinkage requirements, and inflated lead times may all contribute to increased component inventory. Accuracy of the forecast and finished goods safety stock policies, as well as lot sizing considerations, will determine the expected and actual levels of finished goods inventories.

"MRP, if applied conscientiously, can result in fewer shortages with less inventory than an order point (or informal) system. The major operational dependencies are the accuracy of the inventory balances in the system and the accuracy of the bill-of-material information. The bills must reflect all expected usages, including scrap allowances, and must be stated according to the way the product is actually built, as contrasted with the engineering bill or some other theoretical product structure.

"Books on the subject usually specify required inventory accuracy of 95% (some say 98%) and bill-of-material accuracy of 98%+. These are desirable targets, but failure to reach these levels does not mean that you cannot use MRP. It just means that your success with the technique may be limited or you may require more buffers (inventory) to achieve the high availability levels desired. High accuracy and stringently applied procedures will allow you to reduce buffers and manage more closely, achieving the best results. One of the most difficult aspects of MRP implementation is establishing the inventory reporting discipline necessary to achieve the record accuracy required. Good inventory reporting is also the key to validating (and correcting) bills-of-material.

"The other major challenge for MRP users is establishing a reasonable, stable Master Production Schedule (MPS). Because the MRP system is planning the acquisition of all levels of parts and materials to meet the plan as stated in the MPS, any changes to the

MPS within the total lead time (Cumulative Material Lead Time, or CMLT)[2] will have an effect on some on-going activity. The closer to final assembly, the more costly and disruptive changes to the MPS will be. Failure to respect the MPS lead time restrictions will cause expediting, shortages, and production disruptions (changing priorities).

"Developing a Master Schedule involves balancing perceived marketplace demands (forecast or backlog) against the plant's ability to produce the products. If demands exceed production capabilities, there may be a tendency to build a Master Schedule that exceeds the demonstrated capability of the plant. If this happens, the plan cannot be met and customer shipments will fall behind schedule. Developing and maintaining a realistic and stable Master Schedule are the hardest parts of MRP II to implement because a lot of judgment and some difficult business decisions are often involved. The good news is that a well-designed Master Schedule Planning system can assist in the development of the plan and allow the manager to try out various scenarios in the computer and see their effect before committing to a final production plan. As things change, the system can also provide early warning of the impact of changing situations and allow simulation of the impact of various solutions during the decision-making process.

Manufacturing Resource Planning

"MRP II combines MRP with supporting applications in other areas of the company. It is obvious that inventory and standards information are required for MRP. Perhaps it is also clear that management systems to address the execution of the plan in both purchasing and production areas are also necessary parts of a complete system. Less obvious are the applications in other areas of company operations that can also relate to and support the efficient development and execution of the plan. Order Entry, for instance, can capture demand information (which flows through this application) and make it available to the planning applications for

[2] Cumulative Material Lead Time is the total time required to produce an item from start to finish assuming that no materials are available. This includes all manufacturing and purchasing lead times for all components and production stages, considered on a critical-path basis.

use in developing a forecast, comparing the forecast to backlog or actual demand, using this information in building the Master Schedule, and in allocating (reserving) items that have been ordered but not yet shipped so that they are not promised to another customer.

"All activities usually can be related to costs of some sort, so it makes sense to integrate the cost collecting, accounting, and reporting systems into MRP II to avoid duplication of effort and ensure that the accounting system is based in reality. To go back to the Order Entry example, when an item is shipped, this activity can be tied to the invoicing step, which in turn can pass invoice information to receivables and appropriate information to the General Ledger. Integration of these functions helps avoid duplication of effort, introduction of errors, and delays in transferring information from one area to another.

"The additional benefit of integrated applications is the additive effect that results from shared information. Each functional area within the company will take responsibility for the entry and control of the information specific to its function, but will benefit from the availability of information supplied and maintained from other areas. Thus, all benefit by receiving back more than they contribute.

Implementation

"There are more than 200 different MRP II software packages available[3] and most of them include all of the same major functions and operate under similar design. Although there are, of course, some differences, it is difficult to distinguish between the various offerings on function alone. The simple fact is that software deficiency is rarely if ever the cause for lack of success with MRP II.

"There are three major factors that are essential to successful implementation of MRP II. These are leadership, education, and teamwork.

"Leadership must start at the top. Effective use of MRP II, for most companies, represents a major change in the way the business is operated. Such sweeping changes in operating philosophy require the active involvement, full commitment, and absolute support of the top management team. Commitment can flow down, but rarely will

---

[3] The latest tally I'm aware of (1995) lists more than 300 packaged software solutions in the MRP II category.

it flow up. Top management must authorize the allocation of resources required, give direction, encourage, and keep the project a high priority.

"Education is essential to acceptance and effective use of these new management tools. People have a natural fear and resistance to change that only can be overcome by understanding and being comfortable with the system and its function. You will be asking people to make decisions based on what the system presents them. They must know enough about where the information came from to be comfortable in its use and be able to resolve problems. Education and training must extend through all levels. One effect of system implementation is to tie everyone into a single operating plan. All participants should understand their role in the process.

"A short description of MRP II might be: An information control system that helps coordinate all activities to the company goal, which is to produce products on time and in the most effective way possible. Because the primary function of the system is to make information available across functional areas, it is important that the implementation team include representatives of all areas affected. Many of the functions of MRP II include information from and provide information to multiple departments within the company. Decisions made by a single user department without the knowledge and concurrence of all affected areas can result in a less effective system.

Summary

"MRP and MRP II have been with us for a number of years and have been expanded and enhanced continually. The introduction of newer ideas such as just-in-time and Computer-Integrated Manufacturing (CIM) provide us with new paths beyond the basic functions included in MRP II and offer the opportunity to build on what we have gained through the implementation of MRP II systems. All of these techniques and philosophies are being applied worldwide by companies interested in producing quality products, on time, and in tune with changing markets. Effective management of information is a basic building block of modern manufacturing management. MRP and MRP II are the backbone of manufacturing information management."

Despite the fact that the basic idea behind MRP and MRP II has not changed much in over twenty years is both amazing and a tribute to the basic soundness of the approach. Over the last few years, some software vendors have attempted to distinguish their products from the other three-hundred-plus competitors by calling it something other than MRP and/or by trying to disparage the validity of MRP, claiming that the world has changed and MRP is no longer effective.

Market researchers and consultants have also made a number of attempts to define and promote a new acronym to replace MRP (MRP II)—among the candidates are ERP (Enterprise Resource Planning), COMMS (Customer-Oriented Resource Manufacturing Management System), COMS (eliminate the word manufacturing from COMMS), and others. No matter what you call them, and no matter how refined and improved these systems or concepts may be as compared to the original MRP, they are surely direct descendants and not inherently different in their basic approach.

The best argument I've heard for a new name draws an analogy with other technologies—the ancestor of today's airplane was once called a flying machine. We used to call four-wheeled self-propelled vehicles horseless carriages; now they're cars. At what point did these name changes take place? At what point do we stop calling these solutions MRP and adopt a new name?

There have been significant enhancements and improvements to MRP II over its life. In addition to the continuous extension of the applications to more and more of the business functions, these systems are now reaching out beyond the walls of the plant and there have been significant advances in functionality within the bounds of the defined applications.

Comprehensive applications now cover the entire range of a company's business from marketing, design, development, purchasing and production, maintenance and quality management, distribution, and after-sales installation and support. The so-called "enterprise" extension reaches out to other plants or companies within the corporation and beyond to trading partners using Electronic Data Interchange (EDI) and Electronic Funds Transfer (EFT) to automate and integrate.

Internally, some of the basic MRP II processes are now (or soon will be) carried out in new and innovative ways due to the application of ever more powerful processors and other advances in technology. The basic MRP calculation, for example, is no longer necessarily limited to a one-way process (planning down from the top [Master Schedule] to the bottom) and oblivious to resource constraints. The newest planning systems can be finite-capacity-constrained, optimized using artificial-intelligence techniques, and bidirectional—recommending changes to the Master Schedule when an acceptable plan cannot be laid out.

Yes, MRP II has come a long way and promises continued advancement and extension. But, in my view, it is still MRP and I'm very glad we still have it around.

Appendix B
The Future of Packaged Software

The magazine article reproduced here is actually the February 1994 installment of the "Midrange Manufacturing" column that I wrote for *3X/400 Systems Management* magazine and is included here with the permission of Hunter Publishing.

As I stated in Chapter 2, I am a firm supporter of packaged application software solutions despite their limitations. If the changes discussed in this article come to pass, we will end up with the advantages of a package but without many of the disadvantages. In any case, I felt it was worth including this small glimpse into the future of packaged software.

The computer industry is changing, which in itself is nothing new, but the changes taking place at this particular time seem to be more profound than usual. We are entering a new era of computer hardware and operating systems with "Open" systems and "Client/Server" design becoming the dominant features. Similar changes are taking place on the software side with relational databases, fourth-generation languages (4GL), and, especially, object-oriented programming systems.

Article Reprint

Is Packaged Software Dead?

A gentleman named John Porter has written a book *AS/400 Information Engineering* (McGraw-Hill, 1993) that contains, in an appendix, a white paper predicting nothing less than the death of packaged software as we know it. Porter's thesis is that software packages as they exist today suffer the same kinds of fatal limitations as yesterday's mass-production auto factories. Just as "lean production" has changed the paradigm in manufacturing, Porter's so-called "DesignWare" will displace today's fixed function package that Porter calls "ConcreteWare." Packaged software is sold, and purchased, for its functionality: The better the match between package function and company processes, the better and more useful (and successful) the implementation is likely to be. Porter contends that processes are changing as companies reengineer to stay competitive and the only relatively constant aspect of manufacturing information systems is the database. It follows, then, that a solution should be selected for its database (how well it fits the company's needs), but the application functions should be able to change as the processes themselves change.

Porter's vision is that a software vendor will have hundreds of small minimodules of application code, developed with CASE tools and object-oriented design techniques, which can be selected and pasted together as needed to support a customer's specific needs. As the needs change, new or additional modules can be installed and those no longer needed can be removed, or variations can be developed using CASE tools for truly unique needs. These variations then become substitute minimodules that can be easily implemented among existing minimodules.

Compare this to the current situation wherein a packaged software solution is likely to have one inventory management module—one size fits all. If the module is not a good fit, the only options are to live with it as it is or modify the package. With DesignWare, there might be several dozen small minimodules that make up the inventory management function. For a specific company situation, a number of these minimodules are selected and installed together for a more precise fit to company needs. As company needs change, other minimodules can be added or substituted easily, in effect modifying the function while staying within the vendor-sold and supported offerings.

The analogy to manufacturing reaches back to the earliest cars and relates "craft" production techniques to custom software development. Before Henry Ford, cars were fabricated by hand, and each was unique as well as expensive. Similarly, custom-developed software is also unique and expensive since each new program is hand-made from scratch. Mass production introduced the idea of interchangeable parts. Once a part is designed and the tooling prepared, many copies can be produced with relatively little incremental cost. Similarly, with today's packaged software products, once a program is developed and tested, it can be resold many times with little additional cost.

Mass production relies on the so-called "economies of scale" wherein fixed costs are amortized over a large number of copies. But mass production also assumes that consumers will buy what the factories choose to produce. Although a buyer can custom order a particular configuration, he is limited to the choices offered by the car company.

Taking this analogy over to the software side, a packaged solution may offer some choices, such as which modules to include, but the choices are limited to those offered by the vendor. Software packages can be modified, of course, but this takes away much of the benefits of packaged solutions—low cost and centralized support among them. You can also replace the transmission or put a sun roof in your new car, but, again, the economies of scale and advantages of vendor support may be lost.

What's happened in the auto market recently is a proliferation of new models and spin-off brands (Lexus, Infinity), each pursuing a niche market where specific features are tailored to perceived market desires. A more splintered market means lower volume for any specific model, thus erasing some of the opportunities for high-volume production. To make up for the loss of scale, manufacturing plants have had to become more automated and more flexible so that changeover costs can be reduced. In addition, new production techniques called "Lean Production" have evolved to provide increased flexibility and to reduce the economic production quantities. From the customer side, there is more choice and, presumably, a better fit between product and need. The software equivalent to Lean Production is DesignWare, which offers more choices but reduces the volume of sales for any specific module because of the larger assortment available.

The major drawback to the DesignWare idea may be the difficulty in marketing and configuring such a system. Much more upfront analysis will be required in order to select the proper array of minimodules for a specific customer situation. The sales staff will require in-depth business knowledge and manufacturing expertise, as well as detailed knowledge of the operation of each minimodule, to be able to properly configure a solution. If this idea takes hold, I'm sure that there will be software to help in this task, but even with configuration software, the presale activities will certainly be much more involved and more critical than in today's packaged software world.

There will also be a significantly more complicated management task for the hundreds of minimodules that will make up a system, as compared to today's relative handful. Can you imagine the complexity of designing, testing, and delivering two dozen minimodules within an inventory function that can be implemented in any combination? It gives the term "System Test" a whole new meaning.

I don't know whether John Porter's vision will come to pass, but I think he has effectively captured both the limitations of today's packaged software and the general direction of the market. As in so many other areas, the manufacturing software market is becoming more and more fragmented, with literally hundreds of vendors trying to distinguish themselves from the rest of the pack. Today the trend is toward broader function—adding modules for such things as field service tracking, telemarketing support, and management of quality data. This is a limited game because it is relatively easy to add modules and functions. As soon as one vendor introduces a new feature, all of the others quickly follow along. The other major trend today is toward relational database design (virtually all products are already there) and the so-called "Open" and/or "Client/Server" environments, although there is no general agreement on just what these terms mean. Once again, all vendors now have or will soon release products with these characteristics.

It is likely that the next wave of competition will focus on differentiation of product function—a closer fit to the customer's needs. Smaller interchangeable modules offer an attractive path to achieving this goal and the rising popularity of object-orientation programming techniques will lead developers in this direction.

Whether this will all lead to "DesignWare" is an open question. That advances in technology will evolve to support an ever-changing software marketplace and that the software marketplace will evolve to exploit technological advances are a certainty. Other possible solutions exist, of course, and only the passage of time will show us the successful approach.

There is an ancient curse that says: "May you live in interesting times." I choose to believe that it is a blessing to have all of this marvelous technology ready to serve our needs. The sad thing is that far too much of the technology that is available today is not being put to effective use primarily because of a lack of user training and to a lesser extent a failure to organize and manage a sound implementation effort. Technology will advance. Those companies that manage to exploit it will be the winners in the marketplace. The real key to success is not how advanced the tools might be, but how well they are being applied.

Since this article was written, the move toward object-oriented design has marched slowly forward. It is still a case of most software developers expressing interest and proclaiming that a lot of research is going on, but few actual products are yet available in the open market. Object-oriented design has joined Open Systems and Client/Server as buzzwords that appear in the majority of product advertisements and descriptions without significant understanding by the prospective purchasers or real benefit to the users.

To expand just a bit on the ideas contained in the preceding article, when true object designs are available, a vendor will be able to prepackage targeted solutions for specific markets. Vendor X, for example, may now offer one package that it claims can do it all—discrete manufacturing, process, make to order, build to stock, and so on. With objects, as outlined before, this vendor could offer a chemical plant package, a food plant package, an electronics package, automotive package, consumer goods package and so on. Each of these targeted products is a different combination of objects from the entire set (called a class library).

For the user, the result is a closer fit to requirements, along with the advantages of high-volume packaged software. Also, if the customer's needs change or expand, other objects can be "plugged in" to further customize the functionality of the solution, again without custom programming. If custom programming is required, it will be easier to isolate the changes to specific objects, rather than an entire application module, reducing support headaches. It is also likely that customization will be easier to do, and maintain, because of objects' ability to carry characteristics forward through copies of itself (inheritance).

The question that lurks in the back of most managers' minds is: "Should we buy now or wait for the next generation?" There is a fear that if you don't wait, you are buying something that will soon be obsolete. On the other hand, it is risky to be on the leading edge (some call it the bleeding edge) of technology. There is no simple answer. Ideally, it would be nice to buy a system today that is well positioned to take advantage of the new technology as it evolves. I'm not sure that is possible because the coming changes are so basic to system and software design that the new products will not look (structurally and architecturally) much like today's products.

System decisions must be made on a case-by-case basis to consider the urgency of the needs, the adequacy of existing solutions, and the potential advantages of upcoming technologies. There is no single "right" answer, and even if there was one "right" solution today, it may not be the best answer tomorrow.

The only intelligent approach, if you must choose a package today, is to buy from a vendor with the vision and the resources to carry you through the changes. Expect to have to make the transition, and choose a partner that can help you do it when you (and the technology) are ready.

Appendix C
MIS and Systems Implementation

The following is a slightly updated reprint of an article of mine that was originally published in December 1988 in a magazine called *SYSTEMS/3X & AS WORLD* under the title "The Perils of MRP Project Leadership." The magazine has since changed names several times and now goes by the name of *3X/400 Systems Management.* This article is used with its permission.

Note that I used the terms DP (Data Processing) and MIS (Management Information Systems) interchangeably here. Today, the most politically correct department designation is probably IS or I/S for Information Services.

As MIS director of ABC manufacturing company, you are asked to participate in a project to investigate packaged Manufacturing Resource Planning (MRP) software. Since this a data-system project, you become project leader.

After defining the requirements, the search begins. Vendors present their proposals, the team members all review each proposal and you gather their evaluations into a decision matrix and present the results for board approval. Decision made, the order is placed and it's time to think about implementation.

Since you have taken the lead up to this point (and done an admirable job, at that), naturally you are the prime candidate for implementation project leader.

Think twice! This "reward" for a job well done could put you in an impossible position.

Unless your company is in the software business, the MIS or DP department is a *support* organization not a "line" function. It is curious, then, that DP is often tapped to lead an MRP II implementation (Manufacturing Resource Planning system hereafter referred to simply as MRP) since MRP is a business management project that affects all areas of company operations.

Today's MRP systems are designed to be distributed. Each user department has access to data entry, file maintenance, information retrieval, and functional capabilities. Inherent in this design is the requirement that all user functions must take "ownership" of their portion of the system. Materials must take responsibility for inventory transactions and functions, production must be responsible for shop-floor activities, accounting does their own data entry and processing, and so on.

When MIS is put in charge of the project, the system is perceived to be an MIS system. If this is the case, where is the motivation for the stockroom to take the extra time necessary to enter the transactions when it's busy (isn't it always)? It's just too easy to postpone "DP's" data entry when your own requirements just seem so much more immediate and important. The perception of ownership makes all the difference in this situation. It just doesn't work when the users are simply involved rather than being fully committed.

Why do companies tend to assign project leadership responsibilities to MIS? Primarily because they fail to recognize the importance of the "ownership" concept introduced earlier. Because the project involves a computer and the hardware and software represent the most visible up-front costs, MIS is heavily involved in the system selection and it only seems natural for MIS to continue in a leadership role.

Things have progressed to the point where the files are loaded and transactions are being entered. In a status review meeting, it becomes evident that there is a problem with inventory accuracy. The other team members turn to you, the project leader, for the solution. You must point out to them that it is transactions that report activity and result in the incorrect balances. The only way things will get better is if the transaction procedures are tightened.

They seem to not hear or not believe you. Discussions quickly focus on extra reports, system verification of transactions (editing), location of terminals, system security. Because this is a DP project, the solution must be a DP solution.

An MRP system must be viewed as a tool to be used to assist in management decision making. The tool belongs to the department that depends on the tool for job performance. The "system" belongs to the functional departments just as a lathe belongs to production and not to maintenance.

Because MRP crosses functional boundaries, there is bound to be contention on the "ownership" issue. A typical full MRP II system will address production, materials (which may or may not be a part of production), planning and scheduling, customer servicing, finance and accounting, and even engineering, maintenance, and office functions. Since there is no single clear-cut owner, MIS must be the custodian of the hardware and take care of the software-related support (updates, enhancements, and so on). From a functional standpoint, however, MIS is in no position to manage the data entry and maintenance or the day-to-day exercising of the functional activities. Each functional area must take responsibility for its own "piece" of the system, the database, and so on.

MRP provides information management services to the above-named departments. To be effectively used, the data provided to the system must be complete, accurate and timely. The only viable source for the input data is the users themselves. The users are also the only ones who can verify the accuracy of the data. How fortunate that these same users are also the ones who benefit from the management of the data and their conversion into information that they can then use in the management process.

The effectiveness of the MRP system is totally dependent on the confidence that the users have in the validity of the information that it provides. We are asking these people to "bet the ranch" (and their jobs) on this information by making critical business decisions in terms of production schedules, inventory levels, pricing, manpower levels, and so forth. We must provide sufficient education such that the users are comfortable with the process, but equally as important is to place the burden of data integrity squarely in their own control.

You headed the selection committee and now you are in line to be implementation project leader. How can you graciously decline without jeopardizing your status or position?

If the selection process was done right, the selection committee included all major functional areas of the company—engineering, production, materials, finance and accounting, and customer service.

So should the implementation committee (team). MIS played the key role when hardware and software were key issues. Now it's time to let operations take charge.

You will stay a key committee member and supporter. No matter who leads the effort, MIS will continue to have an important part to play.

It is natural to expect that new system users will look to MIS for direction and to solve problems. Embrace this role. Users will ask you to help them get started; they will depend on you (and your staff) during data preparation and loading. When there is a problem, chances are they will call DP. Since you will get the calls anyway, why not become the expert? You can also be a relatively neutral arbitrator if the need arises. Be sure that your role is clearly defined and that the users understand their responsibilities. Don't let them become too dependent on your expertise. Remember that perceptions are extremely important. You will have to resist the temptation to take credit for successes. Spread the credit around.

What about blame? When a computer is involved, it makes an easy target for blame ... after all, it cannot defend itself. Since your position is associated with the machine, you will also be on the hot seat when things don't go as planned. The most important thing is to avoid a finger-pointing contest. Responsibility *must* be established *before* there is opportunity for assignment of blame. You cannot emphasize user responsibility when there is a problem unless the users have been clearly informed and educated up front and conditioned to accept the responsibility. If you don't take a leadership position in educating the users and getting them involved, be prepared to be the scapegoat at a later time.

There are bound to be disagreements and differing priorities. MIS can mediate best if well informed. What better way than to take charge of user education. As education leader or advocate, you can

- educate yourself before and with the users and
- maintain a highly visible leadership position without taking the responsibility that belongs elsewhere.

As soon as the users get a look at the new system, change requests begin to flood in. Everyone wants to make the new system look like the old one as much as possible. You represent a support function and any fears you may have had about being put out of business by the install-

ation of a package are quickly put to rest. Should you immediately start prioritizing and assign people to the project?

The advantages of a packaged software product include

- reduced development cost and time
- reduced support requirements and
- comprehensive function developed to industry standards and practices.

People are naturally resistant to change. It is to be expected that there will be a natural tendency to resist a new approach or a new format because it is unfamiliar. Responding to these requests without question will reduce or eliminate the three advantages just listed.

Changes to a software package are expensive. Programming is a labor-intensive activity. Because the vendor will not be familiar with your changes, he cannot support your modified version of the code. Some vendors will completely cut off support to a modified package; others will only support the unmodified portion of their products. At the very least, support responsibility moves in-house where the expense is borne completely by the user company instead of being shared with all of the other users of the package.

More importantly, duplicating old system (or manual) reports and procedures will only serve to prevent users from taking advantage of new or improved functionality of the package.

Implementation of a new system is a great opportunity to upgrade procedures and practices and to take advantage of improved controls. Modifying the new system to duplicate the old destroys this advantage.

The solution is education. New system users must be comfortable enough with the system to use its information on which to base critical management decisions. Education in this sense also includes a measure of selling and confidence building.

Must all requests be rejected? Of course not. Some may be valid or, indeed, necessary. The best way to sort the wheat from the chaff is to age all change requests. Accept all requests. To reject them immediately will only discourage the users and cause interdepartmental conflict. Review each request after a suitable interval (six months?) and establish whether the request is still valid and if so the cost and possible benefits of the change. Require the justification of each such request on a business case basis, but only after validating the need.

The project is a success. You have contributed to the effort and have avoided the problems associated with MIS control of the MRP project. What now?

It is true that a rising tide lifts all the boats. You have helped your company become more competitive, more efficient, and more profitable. Because you were a key member of the project team, you share in the credit for the accomplishments.

Because you have become more involved in company operations, you can more easily move into operational or general management functions if you so desire. Your experience in this project is great background to be a key participant in follow-on projects to extend MRP into other areas leading to a fully Computer-Integrated Manufacturing (CIM) facility. And remember, an MRP project is not a destination, it is a journey. MRP is never really complete. There is always room for more control, more efficiency, more flexibility. The focus of a just-in-time program, in fact, is the expansion on the basic techniques of MRP to the identification and reduction of wasteful practices and policies.

In summary, it is unfair to ask MIS to be responsible for data and information that they do not produce, collect, or control.

It is unreasonable to ask users and functional managers to make critical management decisions based completely upon information that describes functional realities, is based on data from functional facilities, but for which they are not given complete responsibility.

It is unrealistic to think that "the system" or MIS can provide the answers without each department being fully committed to, not just involved in, the project and the system.

It is unjustifiable to provide such an easy scapegoat (actually, two—MIS and "the system") for failure of the project by letting the users depend on another department to run *their* system.

Index